BEEHIVE ALCHEMY

Inspiring | Educating | Creating | Entertaining

Brimming with creative inspiration, how-to projects, and useful information to enrich your everyday life, Quarto Knows is a favorite destination for those pursuing their interests and passions. Visit our site and dig deeper with our books into your area of interest: Quarto Creates, Quarto Cooks, Quarto Homes, Quarto Lives, Quarto Drives, Quarto Explores, Quarto Gifts, or Quarto Kids.

Quarry Books titles are also available at discount for retail, wholesale, promotional, and bulk purchase. For details, contact the Special Sales Manager by email at specialsales@quarto.com or by mail at The Quarto Group, Attn: Special Sales Manager, 401 Second Avenue North, Suite 310, Minneapolis, MN 55401, USA.

10 9 8 7 6 5 4 3 2 1

ISBN: 978-1-63159-491-5

Digital edition published in 2018

Library of Congress Cataloging-in-Publication Data available

MIX
Paper from responsible sources
FSC® C016973

Design: Burge Agency

Page Layout: Burge Agency

Photography: Kat Schleicher Photography

Printed in China

The information in this book is for educational purposes only. It is not intended to replace the advice of a physician or medical practitioner. Please see your health-care provider before beginning any new health program

BEEHIVE ALCHEMY

Projects and recipes using honey,
beeswax, propolis, and pollen to make
soap, candles, creams, salves, and more

QUARRY

CONTENTS

INTRODUCTION

It was a chance meeting at the local library in 2004 that led my boyfriend Karl and I down the path to beekeeping. On that fateful day, Karl, who had long been fascinated with bees, met a beekeeper's wife who needed help using the library printer. When Karl learned that her husband was a beekeeper, he asked if he ever needed help. She laughed and told him that he could always use a hand—most people were afraid to work with bees. The next week, Karl was helping him, and by the end of the season, we were harvesting honey from our starter hive. Karl was hooked! The next year, we ramped up and got ten more colonies—we also learned that keeping them alive was not all that easy! But we did have some super yummy honey and some awesome white beeswax.

When Karl and I got our first hives and started working with bees, I knew nothing about bees or beekeeping. In my childhood, my mother and I would go to local farms to get eggs or produce,

but we never purchased local honey. I am sure there were beekeepers in the area, but they weren't part of my consciousness. The honey in the house was used medicinally, rather than as an everyday sweetener. We used sugar for all our sweetening needs. So, back in 2004, when we had honey after our first successful honey harvest, I had a lot of learning to do. I immersed myself in the bee world, reading all I could about raising them and using their honey and beeswax in my day-to-day life. At the time, I had no idea that this pantry staple, honey, would be the starting point to a whole new path in life.

Although I knew virtually nothing about bees, I did know a lot about Do-It-Yourself, or DIY as it has come to be known. My love affair with craft started back when I was a child. I am an only child born to German immigrants. My mother, who was trained as a kindergarten teacher in Germany, spent many years teaching in a German school in Peru. It was there that she was inspired to learn to make things by hand and from scratch. She was an excellent cook, accomplished seamstress, and avid gardener. Later, she took up weaving, pottery, and lapidary. She had a mantra that still echoes in my head today: "You can make that!" She was almost always right.

My father had the same do-it-yourself mindset. He was a machinist in Germany and the United States and later worked as an engineer. At home, he was always fixing or making something. In fact, he made much of the furniture in our home when I was growing up. I would often join him to help with his projects and learn about woodworking and mechanical things such as lawn mowers and cars.

Having such inventive parents definitely rubbed off on me, and I've been a do-it-yourselfer all my life. One of my many creative pursuits, a two-year obsession with making soap, has resurfaced since we started keeping bees. Karl quickly discovered that he liked my handmade soaps more than commercially available soaps, so when we started selling our honey at local farmers' markets, we decided to sell my soaps as well. I fine-tuned my recipe and peddled them along with our artisan honeys. It was a great match. With a surplus of beeswax, I started making other items such as lip balms, solid lotions, and salves. My new business, which I called Beehive Alchemy, was born.

Encouraged by the success of my products, I continued adding to my line with creams and lotions, perfumes, and candles. With each product addition, I acquired new ingredients and new knowledge. I also learned the value of experimentation. Following a recipe that comes out of a book or off the Internet can yield a good product, but what makes it good? Having a firm knowledge of the ingredients and why they are in the recipe is extremely important.

Spurred on by the popularity of Beehive Alchemy products, I decided to enter Martha Stewart's American Made competition in 2013. That last-minute decision gave my business exposure to new people and new markets. I also learned that bees were HOT! Everyone wanted my beeswax candles. This visibility also resulted in an offer to write my first book *Beeswax Alchemy*. I poured all my theoretical and practical knowledge of beeswax into those pages.

I never dreamed that four years later, I would be writing this book. My original vision—to help small-scale beekeepers find creative uses for all the products the bees provide—hasn't changed. And, since beeswax is also a big part of this book, there is naturally going to be some overlap, after all, how many ways are there to dip a candle? But as I continue to refine existing products and introduce new ones, I have more knowledge and tips to share. I've included those discoveries and refinements here.

What I love most about this book is the fact that it includes honey, the single most important thing we get from bees. It nourishes us, heals us, and makes most things taste better. Although honey can be used topically for healing and moisturization, it shines in food. During the countless hours I spent researching bees and finding uses for bee products, I came across many recipes that called for honey, which I shared with customers at farmers' markets. Whenever people visiting our market stand set down a jar of honey, claiming that they didn't use much of it at home, I handed them the brochure and told them that they obviously needed new recipes. I'm not sure if it worked on my customers, but it did work for me. With a constant supply of honey on hand, I was always looking for new ways to use it. Replacing sugar with honey in recipes was relatively easy, but I also made a point of learning new transformative ways to use all the bee's products. Although this is not intended to be a cookbook, I've included many of those recipes on the following pages.

CHAPTER 1
ALCHEMY IN THE BEEHIVE

People have had a relationship with bees for many millennia. For most of that time, it consisted of finding the bees in their natural habitat and harvesting the honeycomb. It wasn't until the 1500s that people built structures to house bees and not until the 1800s that moveable frame hives, similar to what are in use today, came to be. These frames allowed the keeper to manipulate the individual combs in a nondestructive way. The 1800s also saw the invention of the bee smoker, the comb foundation maker, and the honey extractor. Although there have been modifications over the years, all of these tools are still being used by today's beekeeper.

HONEY

What we see as the golden nectar of the gods in a jar is in fact made up of an array of different components.

Approximately 80 percent of honey is composed of water and the sugars fructose and glucose. The remaining 20 percent consists of vitamins, minerals, amino acids, enzymes, and polyphenols. The exact makeup of a particular jar of honey will vary significantly depending on the region and on which flowers the bees visited. The wide variations in honey are part of the reason scientists have spent decades researching honey, its constituents, and their effects on our bodies, but there is still a lot that remains a mystery.

HOW HONEY IS MADE

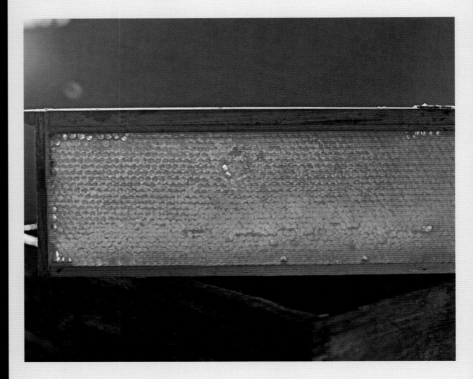

The process of converting nectar into honey starts while the bee is still out foraging. The nectar that a bee collects from a flower is stored in the bee's honey stomach, which is separate from the stomach the bee needs for digestion. While in the honey stomach, the nectar is mixed with various enzymes that start to break down the complex sugars of the nectar and convert it into a more "bee-friendly" simple sugar.

Once the bee returns to the hive, the partially converted nectar is passed to a hive bee to complete the conversion. Once it is converted, the hive bee transfers the nectar to a cell of honeycomb where the nectar evaporates, dropping the moisture content down from almost 80 percent to under 18 percent. This is accomplished in two ways. The first is passive. The ambient temperature in the hive, which is usually between 90°F and 100°F (32°C and 38°C), is hot enough to evaporate some of the moisture. The second way is active. Bees fan the nectar with their wings to evaporate more moisture from the nectar. What remains after this process is what is called *honey*.

HARVESTING AND PROCESSING

Once the cell is full and the moisture content is correct, bees cap off the cell with beeswax to preserve it. When stored in this way, the honey will last indefinitely.

Generally, honey contains some or all of the following:

Vitamins: B6, thiamin, niacin, riboflavin, and pantothenic acid

Minerals: calcium, copper, iron, magnesium, manganese, phosphorus, potassium, sodium, and zinc

18 or more free amino acids

Enzymes and polyphenols

Honey is generally harvested in one of two ways: crushing and straining or through extractor methods. Each method has its advantages and disadvantages.

The crush and strain method is exactly how it sounds. The capped-over honeycombs are removed from the hive, crushed to release the honey, and strained through a bucket top strainer or cloth. This method is great for beginning beekeepers and for top bar hive aficionados because it requires very little equipment—just a bucket and strainer. The disadvantage of this method is that all the wax is removed, which forces bees to make new honeycomb every year. That is an expensive process for bees because it takes 8 pounds (3.6 kg) of honey to produce 1 pound (455 g) of wax. This method also produces a lot of beeswax, which is great for candle makers. There are limitations to how many hives can reasonably be harvested using this method. It's great for the backyard beekeeper, but a bit cumbersome for larger beekeeping operations. This method also leaves more honey trapped in the wax, which could be lost depending on how the wax is purified.

The extractor method involves cutting or scratching the beeswax cap off a traditional Langstroth (or similar design) frame and spinning it in an extractor to remove the honey from the comb. Because the beeswax cap seals the honey in the comb, it needs to be "opened," either by scratching the cap with an uncapping scratcher or fork or cutting off the cap using some sort of knife. The frames are then put in an extractor and the honey is flung from the combs using centripetal force, keeping the main part of the honeycomb intact. The honey flows out the bottom and the cappings are strained to remove the honey. Because most of the comb is preserved, this method is less "costly" for bees in terms of honey needed to produce the wax. The disadvantages to this method are that most extractors are set up to use Langstroth hive frames exclusively, so they will not work with other beehive types, such as top bar hives. Extractors can also get expensive, depending on how many frames they hold and whether they are human powered or electric. This method also produces less wax, so although it's great for the bees, it's not so good for crafters who want lots of beeswax.

BEESWAX

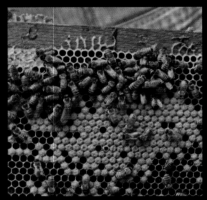

Beeswax is the miracle of the beehive. The comb is built up from nothing and serves as a house, a nursery, and a food pantry. Over the millennia, bees have figured out that building their combs into hexagons allows them to hold the most honey while using the least amount of wax. It also serves as the perfect area for a bee to undergo its metamorphosis from egg to bee.

WHAT IS BEESWAX?

Most basically, beeswax is a wax produced by honey bees of the genus *Apis*. It consists of at least 284 different compounds, mainly a variety of long-chain alkanes, acids, esters, polyesters, and hydroxy esters, but the exact composition of beeswax varies with location. It has a specific gravity of about 0.95 and a melting point of over 140°F (60°C).

More specifically, it is a wax that is secreted from eight wax-producing glands on the worker bee's abdomen. The wax is secreted in thin sheets called *scales*. The scales, when first secreted, looks a bit like mica flakes. They are clear, colorless, tasteless, and very brittle. Beeswax is typically produced by the younger house bees when they are between 12 and 20 days old. As the bee gets older and begins to leave the hive to collect pollen and nectar, these glands start to atrophy, but they don't go away completely. When bees swarm, they will quickly produce wax comb, which serves not only as a place for the queen to lay eggs, but also as somewhere to store food.

HOW BEESWAX IS MADE

To shape the beeswax into honeycomb, the bees line up to form a chain. As the wax-producing bees extrude the wax, it is passed along the chain between the bees' legs and mouths, which are used to chew and mold the wax to shape. The bees will then use this wax to build the hexagon-shaped honey cells we are all familiar with. It is during this process that the wax starts to develop its color and opacity. Depending on what kind of nectar and pollen the bees consumed and what came into the hive, microscopic bits of the pollen and nectar remain and are added to the wax. It takes about 1,100 scales to make one gram of wax.

Under ideal conditions, worker bees can produce beeswax on demand provided there is an adequate supply of food and the right temperature range inside the hive. The ambient temperature in the hive needs to be between 91°F and 97°F (33°C and 36°C). To achieve the correct temperature on cooler spring days, the bees cluster around the wax-producing bees as they are building comb.

The production of beeswax in the hive requires a tremendous amount of resources, however. It takes about 8.4 pounds (3.8 kg) of honey to be converted into one pound (455 g) of beeswax. This honey could be used to feed the nonforaging bees or be saved for periods of nectar droughts. For this reason, bees will often chew off beeswax in one spot and place it where it is needed. The reusing of old comb also contributes to the color, because it may have been used for brood rearing or honey storage and may contain cocoon remains, propolis, or pollen.

Most of the wax that is commercially available is made from something beekeepers call *cappings*. When bees produce honey, the foraging bee collects the nectar and stores it in one of her two stomachs—one is for honey collecting and the other is for personal digestion. The nectar in the honey stomach mixes with enzymes in the stomach, and when the bee returns to the hive, it places the nectar into a waiting cell. As additional nectar is placed into the cells, bees begin to fan their wings to create airflow through the hive to help dry out the nectar. Once enough moisture has evaporated from the nectar to prevent spoilage (less than 19 percent), the bees cap off the cells to prevent additional moisture loss. Bees will systematically work their way across frames or honey boxes, capping each cell as they go.

HARVESTING AND PROCESSING

To harvest the honey, beekeepers remove the frames with honey from the hive and bring them to the honey house for processing. Because all the honey cells are capped with wax, just adding the frames to a honey extractor would yield no honey. So, beekeepers first remove the wax cap using either a hot knife, scratcher, or some sort of flail. The wax cappings then go into a capping tank and the frames go into the extractor where the honey can be spun out.

How beekeepers process the wax cappings depends to a certain degree on how many hives they have. Most systems involve applying heat to the cappings to melt the honey and wax, which will separate into two layers: honey on the bottom and wax on the top. After several filterings, the wax will look pretty clean and is generally ready to use.

Beekeepers also melt down old honey and brood comb in order to install clean wax and do general maintenance on the frames. While wax from cappings and honey combs is fairly pure, the wax from brood combs contains a wide assortment of "stuff," which may include cocoons from both bees and wax moths, excrement from bee larvae, mites, pollen, propolis, and bee parts. All this extra stuff is called *slum gum*. Removing the slum gum from the wax is a more involved process. One method is to put the brood combs into burlap sacks and then add the bag to a hot water bath. The melted wax will flow through the burlap and the slum gum will stay in the bag. The burlap sacks are usually pressed to release the rest of the trapped wax from the

slum gum. The resulting wax is usually significantly darker than the cappings wax, ranging from light brown to almost black. If this wax were to be used for something such as candles, there would still be an unpleasant lingering smell. At this point, a lot of beekeepers turn their wax in to bee supply stores for credit toward *clean* wax, or wax that has already been turned into foundation for inclusion in new frames. The bee supply stores turn this dark wax over to commercial wax processing operations that have specialized equipment with carbon filters to filter the wax and remove the color. This is far better method than in times past, when the wax was literally bleached using noxious chemicals to remove the color. Thankfully, most of the white wax available today is processed naturally using carbon filters, not chemicals.

One drawback to the heavily refined, highly filtered wax is that the aroma and charm of beeswax (as well as many of its unique health advantages) actually come from the natural "contaminants"—honey, propolis, and pollen. Therefore, bleaching or advanced refining of beeswax in order to remove its color and fragrance yields a product that is a bit bland.

PROPOLIS

Propolis is a strange and wonderful thing. More specifically, propolis is the dark brown resinous mixture that bees produce by mixing tree resins, sap, and other botanical elements with saliva and beeswax. Bees use propolis to seal up small cracks and gaps in the hive. They also use it to mummify small animals that die in the hive, covering them in propolis and effectively restoring the hive to sterility.

Propolis is loaded with all sorts of good stuff. It is antibiotic, antiseptic, antiviral, anti-inflammatory, and analgesic. It contains bioflavonoids and polyphenols, which are compounds with antioxidant properties that scavenge free radicals and have an anti-inflammatory effect on the entire body. In comparison to other foods, propolis is one of the highest scoring natural antioxidants available.

HARVESTING AND PROCESSING

Some bees seem to propolize everything. This is a boon for those looking to use the propolis, but can be a nuisance as well. When checking on bees in the summer months, beekeepers often find that everything in the hive is glued in place, from the lid to the boxes and the frames. The propolis needs to be scraped off just to be able to replace boxes and frames. One good thing is that the propolis is soft and malleable because it usually contains more beeswax and at the time of harvest, the temperature outside is warm.

For bees that are a bit more discreet with propolis application, we usually leave most of the propolis in place and only harvest propolis when we are cleaning up equipment. After all the honey has been extracted, we scrape all the boxes and frames to remove the errant wax and propolis. To do this, we set up a makeshift scraping station over a barrel. Any kind of box or container works, but a 55-gallon (208 L) barrel is a nice height for cleaning lots of equipment without an aching back. Propolis can be a bear to work with. What I usually do is tincture the propolis in 95 percent pure grain alcohol. The tincture can then be used either as is or made into propolis oil.

HOW TO USE PROPOLIS

To use the propolis in skin care, I usually take the propolis tincture and create a propolis oil. The propolis oil is easier to handle and can be added to almost anything, just like any other oil.

POLLEN

From each flower they visit, bees collect pollen, which contains the male reproductive cells of those plants. The bees take that pollen and combine it with a bit of nectar and their own digestive enzymes to create pollen granules. They transport the pollen granules by adhering the pollen to their rear legs while they forage.

Much like honey and propolis, the composition of pollen is highly dependent on what flowers the bees visit. On average, the composition is 40 to 60 percent simple sugars, 20 to 60 percent proteins, 3 percent minerals and vitamins, 1 to 30 percent fatty acids, and about 5 percent other stuff. Interestingly, pollen may also contain some fungi and bacteria.

Although the jury is still out on actual health benefits, proponents of bee pollen use it to relieve inflammation, strengthen the immune system, ease symptoms of menopause, and as a dietary supplement.

Beekeepers harvest pollen by placing a screen at the bottom of the beehive, forcing incoming bees to crawl through the screen as they enter the hive. The screen is sized to just fit the bee's body, but not the "baggage." The pollen is shaved off as the bee crawls through the screen and is collected below. Because pollen is actually a source of food for the bees, it is important that the pollen trap only be in place for a couple days, so that the health of the hive isn't impacted significantly.

CHAPTER 2
ALCHEMY FOR THE BODY

In addition to using the honeycomb as a food source, ancient humans learned that the gifts bees provide could also be used externally for wellness. Before the advent of modern medicine, honey was the healing potion of choice. It was used for all sorts of ailments, from ulcers to burns and sores. Honey was the glue for making herbal pastes and ointments. The ancients also realized that beeswax could be used to create healing salves and aromatic unguents. Propolis was recognized early on for its antiseptic qualities. Egyptians used propolis in mummification, and later, Greek and Roman soldiers carried propolis on the battlefield to limit infection. Even pollen plays a role by rejuvenating skin and hair.

With the advent of modern medicine, humans drifted away from the more traditional healing methods in favor of lab-produced antibiotics. Recently, doctors and scientists have rediscovered what the ancients knew: that there is great potential in all the gifts the bees provide.

INGREDIENT PROPERTIES

While all bee-related products may seem to do the same things, it is important to understand that they are interconnected: Beeswax contains pollen and honey, honey contains propolis, propolis contains beeswax, and so on.

HONEY

Honey is a *humectant*, which means it hydrates the skin by retaining moisture. Honey is also a natural source of alpha hydroxy acids, which encourage exfoliation of dead skin cells, refines pores, and smoothes skin. Aging skin loses its ability to retain moisture, so honey can help restore youthful-looking skin.

BEESWAX

Beeswax is a wonderful addition to all sorts of skin care products. In addition to smelling wonderful, it acts as a protective barrier, retaining moisture in the skin and reducing dryness. Not only does beeswax protect the skin, it also allows the skin to breathe and doesn't clog pores. It is a humectant, attracting moisture and keeping the skin hydrated. Beeswax also has anti-inflammatory properties, which encourages wound healing. It is also a good source of Vitamin A, which helps to support cell renewal.

PROPOLIS

Propolis is loaded with vitamins, minerals, phenolic acids, and flavonoids, which means it is packed with antioxidants and nutrients, which ensure that skin remains clear, firm, resilient, and healthy.

POLLEN

Although pollen contains a whole range of awesome ingredients, it's the zinc, vitamins C and E, and the omega fatty acids that are the workhorses for skin care. The zinc acts as an antioxidant that helps slow the aging of the skin. Vitamins C and E protect the skin from damaging UV rays, and omega fatty acids, mostly omega-3 and omega-6, are useful for controlling acne.

SOAP-MAKING EQUIPMENT

Soap-making equipment has only a couple requirements. First off, if using metal, use ONLY stainless steel. The lye will react with other metals. Secondly, never use glass to mix the lye. Over time, minute cracks and scrapes can end up in the glass and the glass will shatter, potentially spilling hot caustic lye water all over. I prefer to keep all the equipment I use for soap making, such as stainless steel bowls and immersion blender, strictly for soap making.

DISPOSABLE PAPER PLATE OR BOWL

For measuring the dry lye, I like to use a disposable paper plate or bowl. This makes cleanup easy. Lye crystals can have quite a bit of static electricity and the individual beads can escape. Take care when measuring and give all the surfaces a quick wipe down when the lye is in water.

IMMERSION BLENDER

For mixing the lye with the oils, I like to use an immersion blender. They can be purchased at a big box store relatively cheaply and they help to ensure that the soap batter is mixed properly. The immersion blender has a high shear blade that combines the soap quickly, easily, and reliably. A hand whisk or spoon can also be used, but it will take quite a bit longer.

STAINLESS STEEL STOCK POTS OR PLASTIC BUCKETS

For mixing the soap, I like stainless steel stock pots or plastic buckets. A small flat-bottomed mop bucket works really well, since it is durable enough to withstand some warmer temperatures and is not too tall, allowing the soap maker to easily use an immersion blender or whisk. It also has a handle, which is useful for transferring the soap batter to the mold.

THERMOMETER

Especially for novice soap makers, it is helpful to know the temperature of the oils and lye water. For this, I like to use an infrared temperature gun. It is noncontact, so it can easily be used to measure both oils and lye water without fear of mixing. It also can be constructed out of whatever material the manufacturer decides because it will not come in contact with lye water. An immersion thermometer can also be used; however, make sure that all the parts that will be submerged in the lye solution are made of a material that is okay to use with lye, such as stainless steel. Also, the thermometer should have a measurable temperature range that includes the suggested temperature of the oils and lye water.

MOLDS

Soap molds can be as simple as a lined cardboard box, as complicated as an ornate silicone mold, or anything in between. The recipes in this book are sized for a regular bread loaf pan (I like silicone, which doesn't need to be lined), which yields approximately 8 to 10 bars of soap.

LINING THE MOLD

For soap molds that require lining, use freezer paper. Place the shiny, coated side up, facing the soap. To line a mold or box, cut 2 pieces of freezer paper and lay them in the box perpendicular to one another, making a crisp corner where the bottom meets the sides and leaving the excess paper folded over the sides of the box. There will be a slight gap at the corners, where the two pieces of freezer paper meet, but I have not found that to be a problem as long as the soap is poured into the mold at the proper time.

SOAP MAKING BASICS

Soap is one of life's basics. Soap has been around since time immemorial, and we all need soap to get us clean. While the actual origin of soap is unclear, people have been combining alkaline salt and fats to create soap in one form or another for the past 5,000 years. The Sumerians created a slurry made by combining animal and vegetable fats with ashes and boiling it. Later, ancient Egyptians wrote about soap recipes that called for mixtures of fats and alkaline salts. The Romans were using bar soap in their baths by 200 CE. What we have now, that they may not have had then, is soap that is reliably gentle on the skin and consistent from one batch to the next. Today, we know about the chemical reaction that takes place to create soap and how to create lye in a lab so that it is of a known purity, ensuring results that can be easily predicted and duplicated. We also have modern tools such as digital scales and immersion blenders that help with the task, but the essential process has not changed in ages.

What is this chemical reaction that happens that creates soap? The chemical process is called *saponification* and at its most basic, it is the following equation:

FAT

Fat in this equation is anything that will react with the lye to create soap. Oil and butters make up the majority of this category, but beeswax falls into this category as well, as a small percentage of beeswax will saponify when introduced to lye.

Each oil has a Saponification Ratio value, also known as the SAP value, which is the amount of lye needed to convert that oil into soap. Eeach oil has a different value, so it is important to be thoughtful about oil substitutions; don't substitute one oil for another without recalculating the amount of lye needed to make the conversion.

LYE

Chemically, lye is either sodium hydroxide or potassium hydroxide, depending on whether a bar soap or liquid soap is desired. Without lye, there is no soap. Lye is required for the chemical reaction. If the soap recipe was calculated correctly, there is no lye left in the final product. All the lye will have been converted to a salt.

The Saponification Equation
Fat + Lye = Soap

CREATING A SOAP RECIPE

Because different oils have different SAP values, they require different amounts of lye. Although it is possible to calculate the amount of lye needed for a particular oil by hand, I recommend using an online lye calculator. I suggest running every recipe through a lye calculator because the recipe could have inadvertent errors such as transposed numbers or incorrect units. It is also a good idea to have a paper copy of the recipe so you can check off materials as you add them and include notes on scent or color. Using the lye calculator also allows for resizing a recipe to fit a particular mold. There are free web-based lye calculators that will calculate the needed lye and allow the end user to print out the recipe. Some calculators have predictive values for hardness and lather that are useful when creating recipes from scratch or tweaking recipes to fit the materials on hand. See appendix C, "Lye Calculations" for the method for calculating the lye quantity by hand.

USING MORE THAN ONE OIL

Although it is possible to make a soap using just one oil, there are reasons that most soap makers utilize more than one. Each oil has a different breakdown of essential fatty acids that will bring different properties to the soap. Coconut oil is high in lauric acid and as a result will yield a soap bar that makes loads of bubbles, will dissolve quickly, and is very cleansing (but may be too harsh for most people if used in high proportion). Olive oil, on the other hand, which is loaded with oleic acid, makes a soap that has more of a creamy lather (it almost feels slimy if not properly aged), is super gentle on the skin, and once fully cured, is hard enough to brick a house. Most of the other oils will fall somewhere in between coconut oil and olive oil. I have included profiles for all the oils I use in this book along with a couple more that might make nice substitutions in appendix D, "Body Care Ingredient Guide."

ADDING BEESWAX AND HONEY

Making soap with honey and beeswax is a great way to utilize some of the gifts from the hive. Since honey is a humectant, it adds a touch of moisturizer to the bar and the sugars in the honey improve the lather.

I consider using honey in soap to be a slightly advanced technique. Although incorporating honey into the batch is not difficult, the honey will cause the batch to heat up a bit more than usual. Here's where experience comes into play; having a sense of the different phases a basic soap batch goes through is really helpful. To use honey in soap, I suggest using about a tablespoon (20 g) of honey per pound (455 g) of oil used in the soap recipe.

Beeswax helps to make the soap emollient and makes a hard bar that lasts longer in the shower. The most common mistake is using too much honey and beeswax, which can lead to a frustrating mess. Use too much honey and the batch can heat up too quickly, possibly causing the mixture to overflow its mold. Too much beeswax and the bar can develop a rubbery and gummy texture. I like to keep my beeswax quantities at 2 percent or less of my oils. Moderation is key.

Beeswax is definitely a more advanced technique. The main reason for this is that beeswax has a melting temperature of at least 145°F (63°C), while most soaps are made at a temperature that is quite a bit cooler, around 100°F (38°C). The oils and lye need to be combined at a higher temperature to ensure that the beeswax remains in liquid form. The higher temperatures will make the soap saponify more quickly, increasing the chances of a failed batch.

Beeswax is not an oil, so not all of the wax will react with the lye. Beeswax contains about 50 percent unsaponifiables. Unsaponifiables are the substances in beeswax that do not react with the lye. The unsaponifiables may be substances that help decrease the trace time in recipes using beeswax, but are also probably the substances that make beeswax nice for the skin.

BAR SOAP WITH HONEY

For the novice soap makers, I suggest first making this recipe without the honey. It is fine to simply leave it out. Learn the process and the phases of soap making first and then try making the recipe with honey.

MATERIALS

Coconut oil (76°F [24°C])	9.3 oz	263.7 gr	35.4%
Rice bran oil	3.0 oz	85.0 gr	11.4%
Olive oil	8.0 oz	226.8 gr	30.4%
Avocado oil	2.0 oz	56.7 gr	7.6%
Shea butter	2.7 oz	76.5 gr	10.3%
Castor oil	1.3 oz	36.9 gr	4.9%
Lye (NaOH)	3.6 oz	103.2 gr	
Distilled water	9.1 oz	259 gr	
Honey	2 oz	57 gr	
Fragrance (if desired)	1 oz	28 gr	

Sturdy, heat-resistant container for lye

Large plastic spoon or high-heat spatula

Container for oils

Stainless steel whisk or immersion blender

Thermometer

Scale

Soap mold (silicone bread loaf pan preferred, but any mold will do)

Freezer paper to line mold, if needed

Yield: Approximately 8 bars of a "standard" size (Actual yield will depend on size and shape of mold used.)

BEFORE BEGINNING

The first thing I like to do when I make soap is to gather all my ingredients together and put them on my table in the order in which they will be used in the recipe. I do this for two reasons. First, I can check to make sure that I have enough of all my ingredients to actually make the soap. It's better to realize that there is no castor oil before making soap, rather than right in the middle of soap making. Second, I can grab the oils in order and check them off on my printed recipe. It helps to ensure that I measure out the correct amounts.

PREPARE THE LYE SOLUTION

1. After gathering the ingredients, make the lye solution, as it will need time to cool. Measure the lye into a disposable paper bowl and set aside.

2. Then, weigh the correct amount of distilled water, reserving 4 ounces (120 ml) in a small container to mix with the honey (if using). Pour the water into the container for the lye. Then, add the dry lye crystals to the water.

Stir until all the lye is completely dissolved. There will be some fumes as the lye dissolves.

The lye mixture will get quite hot, so take precautions and make sure to work on a surface that is heat resistant. Set this mixture aside to cool.

3. If using honey, add the honey to the reserved 4 ounces (120 ml) of water and stir to incorporate. Microwave on high for a couple seconds at a time, if needed, to completely dissolve the honey. Add the honey mixture to the cooled lye mixture. When combined, the mixture will change color and the solution will heat up.

PREPARE THE OILS AND THE MOLD

1. Melt all the solid oils in a heatproof container either on the stove or in the microwave. Once melted, pour the oils into the mixing container. Add the liquid oils and stir to mix.

2. Prepare the mold. If you are lining the mold with freezer paper, do that now. Always make sure that the mold is ready to go before doing any mixing.

3. Check the temperatures of the lye and the oils. Ideally, both should be around 100°F (38°C). If not, warm or cool to temperature.

NOTES:

Always add lye to water, never the other way around. Adding water to lye crystals can cause the lye to erupt out of the container.

Make sure to work in a well-ventilated area and avoid inhaling the fumes.

When using honey, beeswax, or any kind of milk, those ingredients can cause the soap to get hotter than normal and in those cases, it may not be necessary to cover the mold. The recipes in this book contain small amounts of "heating" ingredients and are okay to cover. When making different recipes or using multiple heating ingredients, the reaction may create more heat, and the soap could rise up in spots, creating voids in the soap, or crack, to allow steam to escape. Neither will create an unusable bar; however, the soap may be a bit unsightly.

MAKE THE SOAP

1. Pour the lye/honey water into the oils. With the immersion blender, mix the lye and oils. Notice how it changes from transparent to milky. Keep mixing until it gets to a stage called *trace*. Trace means that when the immersion blender is pulled out of the soap, it leaves a visible trail in the top of the soap. I like to take my soap to a medium to heavy trace, which is almost the consistency of a soft pudding.

2. For novice soapmakers, I recommend not adding any color or scent. When you've made soap a few times and the recipe has become more familiar, this would be the time to add those ingredients. Depending on the temperature of the oil and lye and the speed of the immersion blender, the actual mixing portion should not take that long, maybe five to ten minutes.

3. Pour the soap batter into the prepared mold. Scrape out the soap pot and tap the soap mold on the counter a couple times to make sure there are no air pockets. If needed, smooth out the top with the spatula. Cover the mold with a piece of cardboard to hold in some of the heat.

4. Allow the soap to sit for several hours. Since soap making is an exothermic chemical reaction, the soap will heat up in the next couple hours and transform from what looks like a nice, creamy soap into something resembling petroleum jelly and it will be HOT. That is what is called the *gel* phase. While it's not completely necessary for a batch of soap to go through gel, I think it makes for a better, more consistent, end product.

5. After about 24 hours, the soap should be cool, relatively hard, and ready to unmold and slice. If it still seems a bit soft, leave it in the mold a bit longer and try again after another day or so. Once it seems hard enough, cut the soap into individual bars, keeping in mind that there is still quite a bit of water in the soap that will evaporate over time, which will cause the bar to shrink a bit. Lay out a sheet of freezer paper in an out-of-the-way spot and set the cut bars on the paper, leaving a bit of air space between them. Allow to dry and cure for about 4 weeks.

BAR SOAP WITH BEESWAX AND HONEY

Once the basic recipe has been conquered, it is time to add beeswax.

MATERIALS

Coconut oil (76°f [24°c])	9.30 oz	263.7 gr	34.7%
Rice bran oil	3 oz	85.0 gr	11.2%
Olive oil	8 oz	226.8 gr	29.9%
Avocado oil	2 oz	56.7 gr	7.5%
Shea butter	2.70 oz	76.5 gr	10.1%
Castor oil	1.30 oz	36.9 gr	4.9%
Beeswax	0.50 oz	14.2 gr	1.9%
Distilled water	9.2 oz	261 gr	
Lye (NaOH [sodium hydroxide])	3.7 oz	104.1 gr	
Honey	2 oz	57 gr	
Fragrance (if desired)	1 oz	28 gr	

Sturdy, heat-resistant container for lye

Plastic spoon or high-heat spatula

Container for oils

Stainless steel whisk or immersion blender

Thermometer

Scale

Mold (silicone bread loaf pan preferred, but any mold will do)

Freezer paper to line mold, if needed.

Yield: Approximately 8 bars of a "standard" size (Actual yield will depend on size and shape of mold used.)

1. Gather the materials together and place them on the table in the order in which they will be used in the recipe.

2. Weigh the lye and the water separately, reserving 4 ounces (120 ml) of water to mix with the honey.

3. Add the dry lye crystals to the water. Stir until all the lye is completely dissolved. There will be some fumes as the lye dissolves.

The lye mixture will get quite hot, so take precautions and make sure to work on a surface that is heat resistant. Set this mixture aside to cool.

4. Add the honey to the reserved water and stir to dissolve. If needed, warm the mixture in the microwave for a couple seconds at a time to make the honey dissolve completely. Set aside.

5. If necessary, prepare the soap mold by lining it with freezer paper and keep it close by.

6. Melt all the solid oils and beeswax in a heatproof container either on the stove or in the microwave. Once melted, add the liquid oils and stir to mix. Check to make sure that everything is still melted and nothing has resolidified. If it has, pop it into the microwave until everything is liquid.

9. Pour the soap batter into the prepared mold. Scrape the soap pot and tap the soap mold on the counter a couple times to make sure there are no air pockets. Smooth out the top and cover the mold with a piece of cardboard to hold in some of the heat.

10. After about 24 hours, the soap should be cool, relatively hard, and ready to unmold and slice. If it still seems a bit soft, leave it in the mold a bit longer and let sit for another day or so. Once it seems hard enough, cut the soap into individual bars. Lay out a sheet of freezer paper in an out-of-the-way spot and set the cut bars on the paper, leaving a bit of air space between them. Allow to dry and cure for about 4 weeks, rotating the bars occasionally so that all sides dry evenly.

7. Pour the oils into the mixing container and stir to ensure it is all mixed. Check the temperature of both the lye and the oils. To prevent the beeswax from hardening, the oils will need to be a bit hotter, ideally between 110°F and 120°F (43°C and 49°C). The lye should be right around that temperature as well. Add the reserved honey water to the lye water. It will probably turn colors; mine usually turns some sort of pinkish hue. That's normal.

8. Pour the lye/honey water into the oils, using the immersion blender to combine. Once it is emulsified, but not yet at trace, add the fragrance, if desired. Keep mixing until it gets to the trace phase.

For this soap, I usually mix until it is a light to medium trace. Once trace is achieved, work quickly to get the soap into the mold, in case it begins to solidify.

NOTES:

Always add lye to water, never the other way around. Adding water to lye crystals can cause the lye to erupt out of the container.

Make sure to work in a well-ventilated area and avoid inhaling the fumes.

When using honey, beeswax, or any kind of milk, those ingredients can cause the soap to get hotter than normal and in those cases, it may not be necessary to cover the mold.

LIQUID SOAP WITH HONEY

MATERIALS

Olive oil (pomace grade)	12 oz	340.2 gr	50.0%
Castor oil	4.3 oz	122.5 gr	18.0%
Coconut oil	4.8 oz	136.1 gr	20.0%
Almond oil	1.5 oz	40.8 gr	6.0%
Jojoba oil	1.5 oz	40.8 gr	6.0%
Glycerin	14.6 oz	414 gr	
Potassium hydroxide (KOH [potash])	4.8 oz	136.6 gr	
Distilled water (as needed)	24 oz	700 gr	

Slow cooker

Immersion blender

Medium stainless steel pan

Disposable paper bowl

Yield: About 5 cups (1.2 L) of liquid soap paste

In some respects, making liquid soap is similar to making bar soap, while in other respects, it is totally different. The process of changing fats into soap using a caustic solution is the same, but the actual caustic solution used is different. While liquid soap requires potassium hydroxide (also known as potash), bar soap is made with sodium hydroxide (also known as lye or caustic soda). Technically, they are both salts; however, the salt crystals in bar soap made with sodium hydroxide are crystalized enough to yield an opaque-looking soap. Liquid soap made with potassium hydroxide, on the other hand, doesn't crystalize in the same way and the soap looks more transparent.

When it comes to adding honey to liquid soap, it would seem that simply adding honey to the diluent (usually water) would be the best way to go, but I worry that honey that has not gone through the saponification process might leave the end product susceptible to microbial growth. For this reason, I add my honey before saponification is complete.

I used to make liquid soap the same way I made hot process bar soap, but then I learned about using the glycerin method of liquid soap making. Although the soap doesn't seem to have as long a shelf life as the soap made more traditionally, it is much easier and more foolproof to make. Be aware that the process generates a lot of heat, so I suggest using a metal immersion blender, as a plastic model may warp from the heat.

1. Weigh the oils, add them to the crockpot, and warm them on low heat until completely liquid.

2. Weigh the glycerin and add to a medium stainless steel pan.

3. Weigh the potash and add to the paper bowl. Combine the potash with the glycerin in the stainless steel pan.

4. Heat the glycerin-potash mixture on the stove over low heat until the potash has completely dissolved. This may take a while, since the potassium hydroxide doesn't dissolve easily in glycerin. If a few stubborn pieces of potash remain undissolved, strain them from the mixture before adding it to the oils.

5. Add the potash mixture to the oils in the slow cooker (the temperature doesn't matter). Use the immersion blender to combine. Add the honey and continue mixing until tiny bubbles float to the surface. This usually doesn't take too long, maybe 5 to 10 minutes. Stop mixing, turn off the slow cooker, and cover with the lid. Leave undisturbed overnight.

6. Check the mixture the following day. The paste will be dark, but should look translucent and be quite stiff. It should be fully saponified, but a test dilution will confirm this. To test the soap, weigh out an ounce (28 g) of the soap paste and dilute it with an ounce or two (28 to 60 ml) of distilled water. It may take a while for the paste to dissolve in water, but it will do so on its own with enough time and enough water. Leave the test dilution to sit for a day and come back to it. It should look completely dissolved and without any oil/water separation. If the test passes, the rest of the paste can be diluted.

Once the minimum dilution rate is reached and the soap has a pleasant consistency, apply that dilution rate to the rest of the paste. I like the dilution rate noted in this recipe, but everyone has different preferences.

5

NOTE: To completely dissolve the paste, a certain amount of water is needed (this varies depending on the recipe used). If the soap paste has not completely dissolved by morning, add more water in small amounts until the paste has dissolved completely. Note the total amount of water used. This is the minimum dilution rate needed to dilute the rest of the soap paste.

LIP BALM

The first thing I made once I had my own beeswax was lip balm. Balms may be easy body care products to make, but honestly, they can be the source of a lot of frustration—too hard, too soft, feels waxy, feels oily. I encourage experimentation. Everyone's tastes are different.

A lip balm in a tube needs to be stiff enough to hold its shape, should not melt if kept in a pants pocket, and when applied to lips, it should glide on easily. Pour that same lip balm into a tin, and it will probably be too stiff to be picked up with the finger and applied to the lips. I prefer lip balm in tubes and the following recipe is one of my favorites. I call it basic, but it is anything but—this is an awesome lip balm.

MATERIALS

Virgin coconut oil	1.8 oz	52 g	9.60%
Sweet almond oil	1.8 oz	52 g	9.70%
Castor oil	0.1 oz	2 g	1.00%
Lanolin	0.6 oz	16 g	9.30%
Beeswax	1.9 oz	53 g	30.00%
Vitamin E oil	0.1 oz	0.1 g	0.10%
Rosemary oleoresin extract (ROE)	0.1 oz	0.1 g	0.10%
Lip safe flavor oil (optional)	0.15 oz	0.2 g	0.20%

Knife and cutting board

Disposable paper bowl

Scale

Double boiler or heat-safe bowl

36 lip balm tubes (0.15 oz, or 4 g)

Yield: Approximately 36 lip balm tubes

1. Coarsely chop the beeswax or use beeswax pastilles. Weigh the beeswax, oils, and lanolin and place them directly into the top of a double boiler or heatproof bowl. Gently heat until the beeswax and oils have melted.

2. Remove the melted mixture from the double boiler and add the essential oils and Vitamin E oil. While it is still hot, pour the mixture into the lip balm containers. Allow to cool completely before placing the caps onto the lip balm containers. This recipe makes approximately 6 ounces (170 g) of lip balm, enough to fill 36 lip balm tubes (0.15 ounce, or 4 g, each).

FACE CLEANSING GRAINS

MATERIALS

¹/₄ cup (195 g) uncooked rice (any kind will work)

¹/₄ cup (20 g) ground oats

1 tablespoon (9 g) bee pollen

¹/₄ cup (weight will vary) botanical (optional, see sidebar)

¹/₄ cup (weight will vary) clay (see sidebar)

1 tablespoon (8 g) milk powder (optional, see sidebar)

1 tablespoon (7 g) flax meal

Grinder (preferred) or mortar and pestle

Sieve

Mixing bowl

Yield: Makes approximately ¹/₂ cup (120 ml) powder

Cleansing grains are not a new idea. Asian women have been using ground rice in their beauty regimen for centuries. They can be used to clean and exfoliate, removing dead skin cells, unclogging pores, and improving skin tone. Store the grains as a powder in a small jar or shaker-top bottle. When ready to use, mix single portions with a wetting agent such as water or honey.

The recipe can be tweaked easily to customize the cleanser for different skin types. I have included some suggestions in the "Additions and Substitutions" sidebar. When making the powder, I grind the larger, more dense ingredients first on their own and then again with all other ingredients added. The flax meal creates a gel when hydrated that helps bind the mixture, making it easier to spread across the skin.

1. Grind the rice, oats, pollen, and botanicals (optional) to a powder (in batches if needed). Place the sieve over the mixing bowl and pour the powder through the sieve. Grind any remaining large chunks or discard them.

2. Add the remaining ingredients to the bowl and mix well. Grind the entire batch one more time, pour the finished powder into a jar, and cover with a tight-fitting lid.

3. To use, spoon approximately 1/2 to 1 teaspoon of powder into a small bowl. Add just enough water to turn the mixture into a light paste. Wet your face and apply the paste in circular motion, taking care not to scrub too hard. Rinse off right away or leave on as a mask for a few minutes before rinsing.

ADDITIONS AND SUBSTITUTIONS

The following clays, milks, and botanicals can be added to the basic mixture to customize the grains for a particular skin type. The key is as follows:

A = All Skin Types
D = Dry/Mature
O = Oily/Combination
S = Sensitive

CLAYS

White kaolin clay (S, D)

Bentonite clay (O)

French yellow clay (D, S)

French green clay (O)

Rose clay (A)

MILKS

Non fat dry milk (A)

Buttermilk powder (A)

Goats milk powder (A)

Coconut milk powder (A)

BOTANICALS

Chamomile (D, S)

Rose petals (D)

Lemon balm (O)

Lemon peel (O)

Peppermint leaf (O)

Sage leaf (O)

Lavender buds (A)

HONEY GLYCERIN CLEANSER AND MASK:
TWO METHODS

Honey can be used as a daily cleanser by itself, but it is also the perfect vehicle for adding a few more skin-loving ingredients. I have included a basic recipe and an intermediate recipe for those who feel up to the challenge.

BASIC RECIPE

MATERIALS

Honey	3.2 oz	9 gr	40%
Glycerin	2.8 oz	79 gr	35%
Liquid soap	1.0 oz	57 gr	25%
Lavender essential oil		3 drops	0.07%
Frankincense essential oil		2 drops	0.05%
8-ounce (235 ml) bottle with cap			
Yield: 1 bottle (8 ounces, or 235 ml)			

1. Combine the honey, glycerin, and liquid soap in a small bowl or measuring cup. Mix well.

2. Add the essential oils.

3. Pour into an 8-ounce (235 ml) bottle and cap.

The mixture may be more easily combined if warmed. To maintain as much of the natural enzymes in the honey, combine the glycerin and liquid soap and warm slightly to mix; add the honey once the two are combined and have cooled a bit.

INTERMEDIATE RECIPE

This recipe is my go-to recipe. It is a bit more involved and requires the use of a preservative, but it contains additional ingredients that are wonderful for the face. It also utilizes liquid soap paste (recipe page 27), which can be diluted with various hydrosols, making the cleanser easy to customize for various skin types.

MATERIALS

Liquid soap paste	0.5 oz	14.2 gr	6.59%
Hydrosol (see sidebar)	1 oz	28.3 gr	13.15%
Honey	3.2 oz	91 gr	42.17%
Glycerin	2.8 oz	80 gr	36.90%
Preservative (Optiphen)	0.1 oz	2.3 gr	1.05%
Lavender essential oil		3 drops	0.08%
Frankincense essential oil		2 drops	0.05%
8-ounce (235 ml) bottle with cap			
Yield: 1 bottle (8 ounces, or 235 ml)			

1. Dilute the liquid soap paste with the hydrosol. This may take a while, so I usually do this in a container that can be covered and set aside overnight. By the following day, the liquid soap paste should be fully dissolved.

2. Add the rest of the ingredients and mix well.

3. Pour into an 8-ounce (235 ml) bottle and cap.

HYDROSOLS

I love to use hydrosols. They are the "water" component of an essential oil distillation. The essential oil sits on top of the hydrosol. Often, depending on the plant matter distilled, there is very little, if any, essential oil, such as rose, but there will always be a nice hydrosol. In addition to containing trace amounts of essential oils, hydrosols also contain the water-soluble components of the plant. This makes them functionally different than their essential oil counterparts. When purchasing hydrosols, make sure that the vendor is selling the distillate from the essential oil, not an essential oil dispersed in water. They are not the same thing. Also, because these are not essential oils, the hydrosol may smell a bit different. The following is a list of some more commonly available hydrosols that would work well for use on the face. The key is as follows:

A = All Skin Types
D = Dry/Mature
O = Oily/Combination

Chamomile, German or Roman (A, D)

Frankincense (D)

Geranium (A)

Lavender (A)

Neroli (Orange blossom) (O)

Rose (D)

Rosemary (A, O)

Witch hazel (A)

SOLID LOTION BARS

MATERIALS

Beeswax, yellow	3.1 oz	87 g	30.5%
Coconut oil	0.8 oz	23 g	8.0%
Sweet almond oil	1.5 oz	43 g	15.0%
Jojoba oil	1.6 oz	46 g	16.0%
Cocoa butter	1.2 oz	34 g	12.0%
Mango butter	0.6 oz	17 g	6.0%
Shea butter	1.1 oz	31 g	11.0%
Vitamin E oil	0.1 oz	1 g	0.5%
Rosemary oleoresin extract (ROE)	0.1 oz	1 g	0.4
Scent	0.1 oz	2 g	0.6%
Double boiler			
Molds (enough to hold about 10 ounces [285 ml])			
Yield: Approximately 10 bars (1 ounce, or 28 ml, each)			

Solid lotion bars are an anhydrous blend of butters, oils, and waxes, not emulsified products that contain a large percentage of water. This blend yields a bar that can be easily handled without making a mess, but melts easily on contact to soothe and create an emollient barrier. Although butters such as shea butter and cocoa butter are wonderful for the skin, they do not directly moisturize the skin. The moisture needs to come from other sources, such as the dampness that remains after hand washing. These butters work with the beeswax to create an occlusive barrier to help seal in that moisture. The fatty acids that are contained in the various butters also help to nourish the skin.

Solid lotions, much like lip balms, are very simple in theory, but it can take a while to achieve just the right combination of butters, oils, and wax to match personal preference and regional weather conditions. A person in the tropics, for example, may want something completely different than someone in the Arctic. Experimentation is key. My aim when making a solid lotion bar is to find a balance between emolliency and skin feel. Let's face it, solid lotion bars will leave skin feeling greasier than a lotion, as there is no water to cut the greasy feeling. But with a good recipe, that greasiness can be significantly reduced.

1. Melt the beeswax, coconut oil, sweet almond oil, and jojoba together in a double boiler on low to medium heat until the beeswax is completely liquefied.

2. Add the cocoa butter and mango butter, warm until liquefied, and then remove from the heat.

3. Slowly add small pieces of shea butter to the mixture and allow it to melt.

4. After the shea butter has melted into the mixture, add the Vitamin E oil, ROE, and scent and pour into the molds.

WHIPPED SHEA BODY BUTTER

A body butter is a lusciously thick combination of carrier oils, butters, and sometimes wax. The body butter is formulated to provide intense, long-lasting moisturization for your skin. The beeswax gives it staying power and provides an occlusive layer to keep moisture in the skin. It is best applied after a bath or shower.

TIP: Body butter doesn't need to be whipped, but this recipe would probably be a bit too stiff to apply easily if it wasn't whipped. To modify this recipe so that it can be scooped with fingers without being whipped, try reducing the beeswax by 2 or 3 percent.

MATERIALS

Shea butter	12 oz	340 g	68.0%
Sweet almond oil	2.9 oz	83 g	16.6%
Coconut oil, extra virgin	2.0 oz	57 g	11.4%
Beeswax, yellow	0.6 oz	17 g	3.4%
Scent (if desired)	0.1 oz	3 g	0.6%

Double boiler

Medium stainless steel mixing bowl

Hand mixer or wire whisk

5 jars (4 ounces, or 120 ml, each) with lids

Yield: Fills approximately 5 jars (4 ounces, or 120 ml, each) (depending on the amount of fluff)

1. Melt the shea butter in a double boiler. Add the almond oil, coconut oil, and beeswax. Once everything is melted, transfer the mixture to a medium size stainless steel mixing bowl and use a hand mixer or wire whisk to mix thoroughly. Set the bowl in an ice bath to quickly cool the mixture, stirring with a whisk or blender until the mixture is opaque, yet still somewhat fluid. Set in the freezer for a half hour or so.

2. Remove the body butter from the freezer and beat with the hand mixer to homogenize. Add the scent, if desired, and continue to whip until it reaches a fluffy consistency.

3. Spoon into 5 jars (4 ounces, or 120 ml, each) and cover with tight-fitting lids.

SCENT BLENDS

Natural butters often have an inherent smell. If a scent is desired, choosing essential oils or fragrance that complements the scent of the body butter is best. Here are three essential oil blends that work well with the scent of the butters:

1 g patchouli and 2 g tangerine

2 g lavender and 2 g lemon

1 g lavender, 1 g frankincense, 0.5 g amyris, and 0.5 g helichrysum

HAIR POMADE

Hair pomade is a balm that is formulated to give hair some control and to add moisture. The added clay leaves hair looking natural, not greasy.

MATERIALS

Beeswax, yellow	1.2 oz	34 g	24%
Castor oil	1.5 oz	43 g	30%
Lanolin	0.5 oz	14 g	10%
Sweet almond oil	1.5 oz	43 g	30%
Bentonite clay	0.3 oz	7 g	5%
Scent (if desired)	0.1 oz	1 g	1%
Double boiler			
Wooden spoon			
5 tins (1 ounce, or 28 ml, each)			
Yield: 5 tins (1 ounce, or 28 ml, each or equivalent)			

1. Melt the beeswax, castor oil, and lanolin in a double boiler over very low heat.

2. Remove from the heat and add the almond oil and bentonite clay, stirring well. Add scent, if desired.

3. Pour into 5 tins (1 ounce, or 28 ml, each or equivalent).

BEARD BALM

Beard balm is a leave-in conditioner that moisturizes, conditions, softens and helps style your beard. The shea butter softens and moisturizes, the sweet almond oil conditions, and the beeswax acts as a protectant by sealing in moisture.

MATERIALS

Beeswax, yellow	20 oz	23 g	20%
Shea butter	40 oz	46 g	40%
Almond oil	0.8 oz	23 g	20%
Olive oil	0.8 oz	22 g	19.1%
Cedarwood essential oil (optional)			12 drops
Clary sage essential oil (optional)			10 drops
Double boiler			
Wooden stir stick			
2 tins (2 ounces, or 60 ml, each)			
Yield: 2 tins (2 ounces, or 60 ml, each or equivalent)			

1. Melt the beeswax, shea butter, almond oil, and olive oil in the top of a double boiler.

2. Once everything is liquid, remove from heat and add essential oils, if desired. Stir well.

3. Pour into 2 tins (2 ounces, or 60 ml, each or equivalent).

HERBAL SALVE

Herbal salves are healing ointments that are used for various skin conditions. Generally, they are made from herb-infused oils and beeswax and sometimes essential oils. Herbs for the oil infusions are selected depending on what the desired outcome is—to heal, soothe, relax, cool, and so on. They can be combined to treat chapped hands, wounds, mild burns, bites, stings, rashes, and inflammation.

What kind of oil to use depends a bit on the salve being made and on personal preference. Olive oil is commonly used for salves. It's great for the skin, has a reasonably long shelf life, is readily available, and has a relatively neutral smell. Other oils that work well include grapeseed, sweet almond, and jojoba.

HANDLE HERBS WITH CARE

Making herbal salves from scratch is not hard, but it can take some time. Healing herbs are delicate and require a gentle hand. I like to take the slow method of room temperature infusion to extract all the "goodness" from the herbs. I feel this is the kindest and best way to extract the healing essences. Some people will use heat to speed up the process, and with a watchful eye, this can be done successfully, but it is very easy to scorch the herbs and negate any healing properties.

OIL YIELD

It is a little difficult to know beforehand the quantity of herbs and oil required to yield a specific amount of infused oil. There are many factors that affect the yield, such as size of the herb particles, absorption rate of the oil into the herb, and how well the oil infusion is drained once the infusion is complete. I think generally, there is a 15 to 20 percent loss of oil in the process. I usually make more than what is needed for a specific recipe. Once I have removed the amount required for the recipe, I add a fresh batch of herbs to the remaining oil and start the process all over again.

MAKE THE OIL INFUSION

To make an herbal salve, the first step is to create the oil infusion. This is an inexact science—a handful of this and pinch of that. I like to write my herbal infusion recipes in parts. One part can be any unit of measurement—for example, a handful, a cup, a quarter cup, and so on.

MATERIALS

Herbs (enough to fill jar)

Oil (enough to cover herbs)

Jar large enough to contain herbs (pint canning jars work well for small quantities), lid

Strainer

1. Measure out all the herbs and add them to a glass jar.

2. Then, pour oil over the herbs until they are completely covered. Set the jar aside for 2 to 3 weeks to infuse.

3. When I think of it, I like to take my jar and gently invert it a couple times to "fluff" the herbs and make sure they don't form a clump at the bottom. This is less of a problem with coarse herbs, but for finer or powdered herbs, this is a must. The actual oil yield is also an inexact science. The herbs will absorb some oil and there will be general loss due to oils that stay trapped between the herb material. So, if 3 cups (700 ml) of oil went into the jar, perhaps only 2½ cups (570 ml) will be strained out. Once the herbs have infused the carrier oil with all their goodness, it's time to strain the oil and make the salve.

MAKE THE SALVE

MATERIALS

Herb-infused oil	3.3 oz	95 g	79.0%
Beeswax	0.8 oz	24 g	20.0%
Vitamin E oil	0.05 oz	1 g	1.0%

Essential oils (optional)

Double boiler

Wooden stir stick

4 tins or jars (1 ounce, or 28 ml, each)

Yield: 4 tins (1 ounce, or 28 ml, each)

1. Heat the infused oil and beeswax in a double boiler until the beeswax is completely melted. Take care not to overheat the oil or you may risk it breaking down.

2. Once the mixture is completely melted, remove from heat and add the Vitamin E oil and essential oils, if desired.

3. Stir to mix well and pour into prepared containers. If using plastic jars, let the mixture cool before pouring into the jars, as they can warp with too much heat.

NOTE: If Vitamin E oil was added to the infusion before storing it, there is no need to add additional oil.

SALVE IDEAS

CALMING SALVE (PERFECT AT BEDTIME)

1 part chamomile

1 part rose petals

SUMMER SALVE (GREAT FOR BUG BITES, STINGS, BURNS, AND SCRAPES)

1 part plantain leaf

1 part comfrey leaf or root

1 part burdock root

HEADACHE BALM

3 parts peppermint leaf

1 part lavender

1 part eucalyptus leaf

TRAUMA SALVE

2 parts Arnica blossoms

1 part St John's Wort

PROPOLIS SALVE

I don't consider making propolis tincture an exact science. I don't worry too much about grinding it or breaking it up. Most of my propolis is hive scrapings, so I'm not dealing with big balls of propolis. If they are bigger chunks, it might be a good idea to break them up somewhat. Of course, if it's pure propolis, it should all dissolve anyway.

Propolis salve is a different animal from an herbal salve. Propolis doesn't easily dissolve into oil, so it requires a couple steps to make the salve work. I have seen recipes that call for grinding up propolis and adding the powder to an oil/beeswax mixture. For me, the problem with this approach is twofold. First, my propolis contains a lot of "other stuff," mostly beeswax. Perhaps our bees aren't as "careful" as they should be, but I suspect other beekeepers may have the same problem. Simply grinding up our hive scrapings would probably not be as useful nor as beneficial as straight propolis.

Second, the act of grinding the propolis to a powder generates heat. Even if the propolis starts out frozen, it won't remain frozen through the grinding process. Even if grinding were possible, mixing a powder with a salve will still add an unpleasant grittiness.

I have a better solution, although it requires some time. The first step is to make a propolis tincture.

PREPARE THE PROPOLIS TINCTURE

MATERIALS

Hive scrapings

95% pure grain alcohol

Large plastic container with a tight-fitting lid

Paper coffee filter

Jar large enough to hold hive scrapings, lid

1. Add raw propolis to a large plastic container with a tight-fitting lid.

2. Pour in enough grain alcohol to completely submerge the propolis and cover with the lid.

3. Set it aside and shake it occasionally to mix and break apart chunks.

4. After 1 week or more, strain the liquid through a coffee filter into a glass jar.

The propolis tincture is now ready to use as is or to be made into propolis oil.

PREPARE THE PROPOLIS OIL

MATERIALS

Propolis tincture

Castor oil

Wide-mouth jar, or shallow container large enough to hold the dry herbs

1. Take the filtered propolis tincture and add an equal amount of castor oil.

2. Stir it to incorporate and leave the lid off the container to allow the alcohol to evaporate. I usually allow about 2 weeks for the alcohol to dissipate, stirring occasionally. The oil should be a clear, dark yellow/amber color. All the propolis should still be suspended in the castor oil.

The propolis oil is now ready to be used as is and can either be applied directly to the skin or made into a salve.

MAKE THE SALVE

MATERIALS

Beeswax	0.8 oz	23.8 g	27.0%
Olive oil	0.5 oz	15.5 g	17.6%
Almond oil	0.6 oz	15.9 g	18.1%
Grapeseed oil	0.5 oz	15.5 g	17.9%
Propolis castor oil	0.5 oz	13.5 g	15.3%
Vitamin E oil	0.1 oz	4.0 g	4.5%
Double boiler			
Wooden spoon			
3 tins (1 ounce, or 28 ml each)			

Yield: Makes approximately 3 ounces (90 ml), enough for 3 tins (1 ounce, or 28 ml, each)

1. Heat the beeswax with the olive, almond, and grapeseed oils in a double boiler until the beeswax is completely melted.

2. Remove from the heat and add the Vitamin E oil and the propolis castor oil. Stir to mix well.

3. Pour into 3 tins (1 ounce, or 28 ml, each).

OLIVE AND HONEY LOTION

Learning to make lotions and creams with honey and beeswax was a complicated journey for me. The honey was no problem because it is only a humectant in lotion and not a critical part of the emulsification system. However, the beeswax was problematic. In the beginning, I was resolute in trying to make things work with beeswax. Beeswax by itself is not an emulsifier, but it can be used together with borax (sodium tetraborate) as an emulsifying system. I was able to achieve one single batch of lotion that stayed emulsified, but all the rest either never came together as an emulsified product or they separated within hours. Upon further research, I learned that it is a very difficult product to make successfully using the tools readily available to a home crafter. Commercial operations use fancy homogenizers to mix the materials at very specific temperatures to achieve a product that looks and feels good and is stable, meaning it will not separate over time.

Making lotions and creams are more advanced projects. With balms and salves, end products can be melted and adjusted until a suitable result is achieved and there is no risk of bacterial contamination because they contain no water. Lotions and creams, on the other hand, require more precision. The ingredients are broken up into phases that are heated and combined separately before the phases are combined to make the lotion. Temperature and precise weighing of ingredients is critical to success, as is proper protocol for ensuring sterility throughout the process.

Begin by sterilizing all equipment that will touch the lotion or cream (bowls, spatulas, immersion blender, and so forth) with boiling water or alcohol. I usually use a mix of both. I heat up a pot of tap water to boiling and rinse my bowls and utensils with the boiling hot water and set aside to cool. Then, once I have all my lotion-making ingredients assembled, I give all the equipment a quick wipe-down with isopropyl alcohol (approximately 91 percent alcohol). The alcohol evaporates quickly, leaving a clean, sterile surface.

This is my favorite year-round lotion. It works well from head to toe. It is also a great starting point for creating other lotion recipes. By substituting some of the oils in Phase B, this basic recipe can be modified for specific areas, such as using borage or meadowfoam oils for a face lotion.

MATERIALS

Phase A (Water Phase):

Distilled water	11.0 oz	312.0 g	67.8%
Honey	0.8 oz	23.0 g	5%

Phase B (Oil Phase):

Olive oil	2.6 oz	73.6 g	16%
Shea butter	0.6 oz	18.4 g	4%
E-wax NF	0.8 oz	23.0 g	5%

Phase C (Cool Down Phase):

Citric acid	0.1 oz	2.3 g	0.5%
Preservative (optiphen)	0.2 oz	4.6 g	1%
Scent (optional)	0.1 oz	3.5 g	0.75%

2 small stainless steel bowls or pans

Double boiler

Immersion blender

Noncontact infrared thermometer

Spatula to scrape bowls

4 bottles (4 ounces, or 120 ml, each) with lids or pump closures

Yield: 4 bottles (4 ounces, or 120 ml, each)

1. Sterilize all equipment that will come in contact with the lotion ingredients.

2. In separate pans, *heat and hold* the ingredients for Phase A and Phase B to a temperature of 158°F (70°C); maintain the temperature for 20 minutes.

3. Pour Phase B into Phase A and combine, using an immersion blender on high.

4. Set the container in an ice-water bath to cool it down quickly, continuing to use the blender on high.

5. When the mixture reaches a temperature of 104°F (40°C), add Phase C. Ingredients for Phase C can be premixed or added separately to the lotion. Continue mixing on low until Phase C has been thoroughly incorporated, taking care not to add any air.

6. Pour into 4 bottles (4 ounces, or 120 ml, each).

Heat and hold is a process that holds the phases at 158°F (70°C) for 20 minutes so that the water phase is hot long enough to kill off whatever pathogens might be in the water and to completely melt all the particles in the oil phase. This can be accomplished in a double boiler.

BEESWAX CREAM

In my journey to use beeswax in my emulsified lotions and creams, I finally landed on a compromise. I would utilize beeswax as a thickener, rather than relying on the beeswax to help with the emulsification system. The following recipe is a good starting point and makes a great all-around body cream with a nice skin feel that absorbs quickly and doesn't feel too greasy. For more specialized lotions, oils and butters in Phase B can be swapped out to make a cream that achieves a particular end result. If the same consistency is desired, try to keep the percentage of butters the same and the percentage of oils the same. Do not change the proportion of emulsifier to oils and wax.

MATERIALS

Phase A:

Water, distilled	9.2 oz	260 g	56.5%
Honey	0.3 oz	9.2 g	2%

Phase B:

E-wax NF	0.8 oz	23 g	5%
Stearic acid	0.8 oz	23 g	5%
Beeswax, yellow	0.8 oz	23 g	5%
Mango butter	0.8 oz	23 g	5%
Coconut oil	2.4	69 g	15%
Lanolin	0.3 oz	9.2 g	2%
Vitamin E	0.5 oz	13.8 g	3%

Phase C:

Preservative (optional)	0.2 oz	4.6 g	1%
Scent (optional)	0.1 oz	2.3 g	0.5%

2 small stainless steel bowls or pans

Double boiler

Immersion blender

Noncontact infrared thermometer

Spatula to scrape bowls

4 jars (4 ounces, or 120 ml, each)

Yield: 4 jars (4 ounces, or 120 ml, each)

1. Sterilize all equipment that will come in contact with the cream.

2. In separate containers, heat and hold Phase A and Phase B at 158°F (70°C) for 20 minutes. Pour Phase B into Phase A and combine, using an immersion blender on high.

3. Place the container in an ice-water bath to cool the mixture down quickly, continuing to use the blender on high.

4. When the mixture reaches 104°F (40°C), add Phase C. Phase C items can be added separately to the cream or premixed and added all at once. Continue mixing on low until Phase C has been thoroughly incorporated, taking care not to add any air.

5. Pour into 4 jars (4 ounces, or 120 ml, each).

PROPOLIS TOOTHPASTE

One of the things that propolis does well is counteracting elements that cause tooth decay. Tooth decay is caused by bacteria that convert to acids when sugar is present. Propolis is known for inhibiting the growth of this bacteria.

MATERIALS

1/4 cup (56 g) coconut oil, melted

1/3 to 1/2 cup (74 to 111 g) baking soda

1/4 teaspoon kaolin clay

1/4 teaspoon myrrh powder

1/4 teaspoon stevia powder

2 teaspoons propolis tincture

30 drops of peppermint essential oil

Small pot or microwave-safe container

Small mixing bowl

Spoon

Yield: Fills approximately 2 jars (2 ounces, or 60 ml, each)

1. Heat the coconut oil in a microwave using a microwave-safe container or in small pot on the stove until melted.

2. Combine the baking soda, clay, myrrh, and stevia powder in a small mixing bowl.

3. Pour the melted coconut oil into the bowl and mix well. Add the tinctures and essential oil and combine. Stir until the coconut oil cools, forming a nice thick paste.

4. Spoon into 2 jars (2 ounces, or 60 ml, each).

5. To use, scoop out a small amount of paste and apply to a toothbrush.

HONEY PROPOLIS LOZENGES

MATERIALS

- $1/2$ ounce ginger root
- $1/2$ ounce peppermint leaf
- $1/2$ ounce elderberries
- $1/2$ ounce lemon verbena
- $1/4$ ounce licorice root
- $1/4$ ounce wild cherry bark
- $1/4$ ounce marshmallow root
- $1/4$ ounce horehound
- 2 cups (475 ml) water
- $1 1/2$ cups (510 g) honey
- 1 ounce (28 ml) propolis tincture
- 10 drops of peppermint oil
- 5 drops of lemon oil
- 5 drops of rosemary oil
- Marshmallow root powder, slippery elm powder, or tapioca starch (for dusting the lozenges to keep them from sticking together)
- Large pot with a thick bottom and lid
- Large spoon
- Strainer candy thermometer
- Silicone-lined baking sheet or candy molds
- Wax paper
- Airtight jar with lid
- Yield: Dependent on lozenge size, but volume is about 2 cups (475 ml)

Although I don't get sick very often, when I do, it hits me like a Mack truck. If I have a cough, I go looking for the cough drops and usually find they have been eaten. Stuck in the house with no relief, I have learned to make my own lozenges. I am calling these lozenges, instead of cough drops, because besides honey, they are loaded with herbs, which can be tweaked to personal preference, availability, and need. Propolis's mild analgesic effect is a welcome relief for sore throats.

It is possible to spoon these out onto a silicone-lined baking sheet, but I prefer to use candy/chocolate molds for this. The mixture is fluid enough that it pours easily from the pot and spreads on its own to fill the molds.

1. First, make an herbal decoction by adding all the herbs and water to a pot and bringing it to a boil. Reduce the heat, cover, and simmer for 30 to 45 minutes. Remove the lid and continue simmering to reduce the volume to 1 cup (235 ml) of liquid. Strain.

2. Add the honey and the tincture to the herbal decoction and cook on medium heat. Stir often and keep a watchful eye on it. Cook until it reaches the *hard-crack* stage (300°F, or 150°C).

3. Add the essential oils just as the syrup reaches 300°F (150°C).

4. Pour into candy molds or drop by the spoonful onto the silicone-lined baking sheet.

5. Once the drops have cooled, immediately dust them with something that will keep them from sticking together. Marshmallow root powder is great for this, but even tapioca starch would work.

6. Roll them in the powder, wrap them in wax paper, and store in an airtight jar. Best enjoyed within 2 months.

CHAPTER 3
ALCHEMY OF LIGHT

Candles made of beeswax were used in ancient Egypt, Greece, Rome, and China, perhaps as early as 3000 BCE. Ancient candles were basically paper rolled into a narrow tube and dipped into beeswax or tallow. Today, we know more about the science of candle making. Beeswax candles are more than wax and a cotton string. They are a symbiotic relationship between, wax, wick, and oxygen.

WAX

I'm not crazy about rendering wax, so I let Karl handle the "heavy lifting" of rendering the cappings into big blocks of wax. The wax that Karl renders out is really pretty clean, but since my candle business has outgrown what our hives can produce, I also purchase wax from another beekeeper in the area. His wax varies from relatively clean to blocks with rivers of honey buried inside.

For making candles, especially pillar candles, the residual honey in the wax causes the wax to burn unevenly and to clog the wick. Even though a wax may look clean, it may still have honey in it. When the honey heats up, it caramelizes and forms dark flecks in the wax. Besides being unsightly, these flecks clog the wick and ultimately, keep the wax from reaching the flame.

The best way to get the last of the honey out of the wax is to allow it to clarify in a heated double boiler or wax tank. Admittedly, this task is easier to accomplish with the wax tank than a double boiler because the wax needs to remain as a liquid for quite a while until all the honey has settled to the bottom. I usually let mine settle for a couple days. The best way to tell if it is done is by checking the clarity of the wax. When it is first melted, it has murkiness to it. As the honey settles out, it begins to clarify. When the wax is clear, filter the wax through a clean piece of felt cloth and mold into useable portions. I usually do a variety of different sizes, so that I have the right size for whatever I am making. The resulting wax is still yellow and still has the signature honey-like scent, although the filtering may have lightened up the wax a little bit.

TEMPERATURE

When I first started making candles, I used the melted wax at whatever temperature it was when it came out of my wax tank. Sometimes, the candles worked beautifully, and other times, they failed miserably. In an effort to achieve some consistency, I bought a noncontact thermometer gun and started measuring the wax temperature right before I poured my candles. All of a sudden, my candle success rate went up significantly. I found that if I stirred the wax and allowed the wax to cool to the right temperature, I had fewer blowouts in making pillar candles and my molded tapers came out of the molds blemish free. The actual temperature needed for the different candle types will vary depending on wax, candle mold material, and environment.

STRUCTURE OF A CANDLE FLAME
temperature 1,832°F (1,000°C)

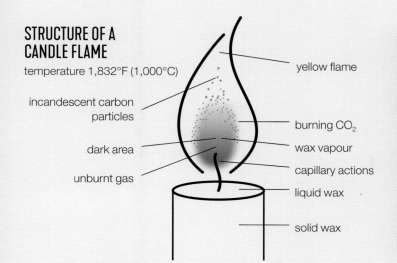

incandescent carbon particles

dark area

unburnt gas

yellow flame

burning CO_2

wax vapour

capillary actions

liquid wax

solid wax

The heat of the flame converts the liquid wax to vapor which travels up the wick and combines with oxygen in the air to produce light, heat, water vapor (H_2O) and carbon dioxide (CO_2). When your wax is clear and cooled at the proper temperature, you've chosen a quality wick, and your candle is designed to ensure the resulting flame is receiving the right amount of oxygen to sustain it, the result is almost uncanny.

WICKING

Although the graph below is not entirely correct in terms of actual diameter of wicking, it gives an overall picture of relative sizes, commonly available wick sizes, and the range of uses. The design and nomenclature of this wicking is, I believe, somewhat unique to the United States. Apparently, wicking produced and sold in other parts of the world utilize a different grading system and are not the same.

I can't stress enough how important the wick is to the overall quality of the candle. This key feature of the candle seems as if it would be the easy part, but honestly, it can be maddening to figure out. When a candle is lit, a series of events take place. First, the match lights the wick and the wick itself begins to burn. The flame then starts to melt the wax. The wick acts as a pipeline that carries the melted wax in the form of a vapor to the flame via capillary action. Some wicks allow lots of fuel to flow quickly through a big pipe, while other wicks pump fuel more slowly through a smaller pipe. If you give the flame too much or too little fuel, it will burn poorly or sputter out. The balance of fuel and flow needs to be just right.

Wicks are made of braided cotton and the nomenclature of square braid cotton wicking refers to the number of bundles, the ply of the wick, and how tightly it is braided. The #6/0 to #1/0 range of wicks are constructed a bit differently than the larger wicks, but all of them are square, which helps to channel the wax fumes up to the flame. It is important to keep all the wicks well labeled and separated because similar sizes look identical. Often the only difference is the tightness of the braiding. Square braid wick forms a carbon cap on the top of the wick. The carbon cap radiates heat outward from the flame, helping melt wax, which is farther away from the flame. The wick also bends slightly as it burns, minimizes carbon buildup and making for a cleaner burning candle.

SQUARE BRAID WICK SIZES

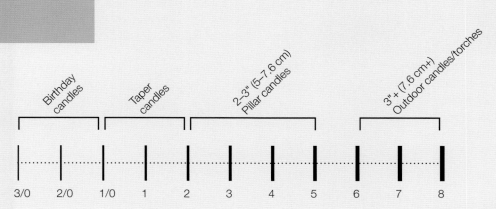

Birthday candles

Taper candles

2–3" (5–7.6 cm) Pillar candles

3" + (7.6 cm+) Outdoor candles/torches

3/0 2/0 1/0 1 2 3 4 5 6 7 8

OXYGEN

The oxygen would seem to be the easy part—either the flame gets oxygen or it doesn't. But the type of candle and the environment in which the candle is burned play a role in how much oxygen the flame receives. I have found that the more open to air the flame is, the better the candle burned. Taper candles are perfectly set up for this. Pillars and votive candles, on the other hand, typically begin burning beautifully, but as they burn wax and the flame travels down into the candle, the flame struggles to remain lit. Either the flame tunnels down the middle of the candle, melting very little wax and starving the flame of oxygen, or the flame melts too much of the wax, which floods the wick and extinguishes the flame.

So, how does one ensure that the candle flame gets the right amount of oxygen to sustain it? Look at the burn pool. The width of the tunnel created by the burn pool is usually determined by the initial burning of the candle. The burn pool, which is an expanse of melted wax on the candle's surface, establishes the ultimate diameter of useable wax available to fuel the candle during subsequent burnings. The solid wax remaining around the outside of the burn pool will help the candle retain its shape. For this reason, I always tell my customers that beeswax candles are intended to be burned all evening, not just for a couple minutes and then extinguished. The combination of proper burning protocol and correct wick size should ensure that the burn pool reaches the desired width.

BASIC BURN TEST

Beeswax can vary from batch to batch. It is best to conduct burn tests with new batches to ensure that you are using the correct size wick. Also, be sure to conduct burn tests when trying a new candle size.

1. Trim the wick to 1/4 inch (6 mm). Make sure the candles are clearly labeled if more than one wick is being tested.

2. Place the test candles in a draft-free spot that is clearly visible from the workspace.

3. Do not leave lit candles unattended.

4. Place the candles on a clean, flat, heat-resistant surface 3 to 6 inches (7.5 to 15 cm) apart.

5. Light the candles and note the time. It is critical to keep an eye on the candles while they are burning, especially when testing new wicks.

6. When testing pillar candles, burn for 2 hours and then record the details of the melt pool and wick appearance. Ideally, the melt pool will achieve the desired diameter within the two-hour time frame. If it hasn't, the wick is most likely too small. Note any soot or mushrooming on the wick.

7. Continue burning for another 2 hours. Record the details of the melt pool and wick again after 4 hours and then gently blow out the flame. At this point, the melt pool of a well-wicked candle will have achieved the desired diameter and should be approximately 1/2 inch (1 cm) deep.

If the wick is mushrooming, the candle is sooting, or the melt pool is substantially deeper than 1/2 inch (1 cm), the wick is most likely too large.

8. Allow the candle to cool for at least five hours and repeat steps 4, 5, and 6 until the candle is completely burned. The quality of burn will almost always change throughout the span of the burn test. Burn the entire candle before making a final decision about changing the wick.

Here's an example of the perfect burn.

MOLDS

Candle molds come in a huge variety of shapes, sizes, and materials. There are pros and cons to all of them, so I will share my own experiences to help with the decision-making process.

TIN MOLDS are some of the oldest molds that are still in use today. They are thin sheets of tin molded into shape and welded at the seams. The one problem I have with these molds involves the seams. Although the finished candle looks good and the molds work well, the seam is visible in the round molds. The rectangular molds, on the other hand, are perfectly suited, since the seam is hidden at the corner. One of the things I like about these molds is that a base is usually welded on as well. The base allows room for the wick to extend through the bottom of the mold, enabling the candle to sit perfectly upright and stable.

ALUMINUM MOLDS are molded rather than formed, which means that there are no seams. Because of this, all sorts of shapes are possible when using aluminum molds that can't be made with tin. My favorite feature of these molds is that they have a molded chamfer around the top edge that gives the candle a nice finished look. Unlike the tin molds, the aluminum molds are formed without a base. With the wick sticking out the bottom, the candles will wobble a bit and lean ever so slightly.

SILICONE RUBBER MOLDS are the newest molds available. They can be molded into almost any shape and are flexible enough to allow for easy removal of the finished candle. These molds have the thickest side walls, which affect how the wax cools. I find that the silicone holds on to heat a lot longer, and for the larger diameter candles, this can pose a bit of a problem because the wax isn't able to shrink as easily while cooling. I prefer this material for my taper molds, where the diameter is less than an inch (2.5 cm) and all the shrinkage can be accomplished in the height.

COLOR

Sometimes, it's nice to add some color to the candles. With beeswax candles, this can be challenging on several levels. The first involves color theory, since beeswax is usually some shade of yellow. That yellow tint combined with red makes orange, with blue makes green, and with black makes olive green. Subtle colors, such as light blue, for instance, can be difficult, if not impossible, to achieve. The yellow mixes with the blue to create a green shade. Adding additional color to make a darker blue will yield a candle that has a greenish tinge to it.

To counteract some of this color shift, I often mix our own yellow wax with some purchased white beeswax. I think of yellow beeswax as being a yellow-pigmented wax, whereas white beeswax is a pigment-less wax rather than white-pigmented wax. So, by blending the yellow wax with white, it is only diluting the yellow color, not changing it.

Candle dyes are what I use to color my beeswax candles. Candle dyes come in a variety of forms, including concentrated color blocks and chips, liquid dyes, and pigment dyes. I have worked with only the liquid dyes because they seemed the easiest to use and replicate. Liquid dyes are highly concentrated, which makes them very economical, but I have found that sometimes I prefer a very light color. For those times, I dilute my dyes with a vegetable or mineral oil.

I have been asked about using natural colorants in candles. Perhaps there are some that have been able to do this successfully, but I have not. Usually, the natural colorants are simply powders that are suspended in wax. Those particles then end up clogging the wick, which leads to a poorly performing candle. Also, some things are just not designed to be burned. Just because it is natural, does not mean it won't be harmful once burned. I stick to the candle colorants that are specifically designed for candles; these colorants won't clog a wick and are made to be burned.

HAND DIPPED CANDLES

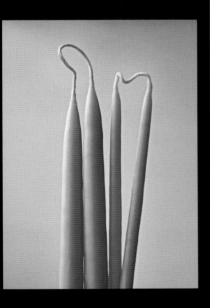

Hand dipped candles have been around forever, probably because they are easy to make. A leisurely afternoon can yield a couple dozen passable tapers, especially if they don't have to be perfect. If the perfect taper is the goal, however, there is a bit more craft involved. I will first go through the basics, and then I will follow up with more details for how to achieve nearly perfect, beautiful candles.

Hand dipped taper candles are made from paper-thin sheets of wax that are added one layer at a time by dipping the wick repeatedly in hot wax. Most regular taper candles have 20 or more layers of wax. To achieve the best results, keep the following key points in mind as you prepare to make hand dipped tapers.

FIND THE RIGHT CONTAINER

To make hand dipped candles, access to a tall wax-melting container is essential. The length of the finished candle is limited by the depth of the wax in the dipping container. A coffee can, a container that many people already have around the house, works great. The only disadvantage to using it is that it will probably only yield a 3- to 4-inch (7.5 to 10 cm)-long candle. To make taller candles, look for a tall, narrow metal container. Something such as an asparagus pot works well. There are commercially available dipping pots that are really tall, and I used one early on, but then the challenge became keeping the wax evenly hot from the bottom to the top. This requires the use of an outer pot for the water bath that is also quite high. It doesn't need to be as tall as the dipping pot, but the water level should reach at least two-thirds the level of the wax to ensure that the wax is hot from top to bottom.

HAVE A GOOD SUPPLY OF WAX

Also make sure there is plenty of wax on hand. Every time the wick is dipped in the wax, wax is removed from the pot. Once a few candles have been dipped, the subsequent candles will not be as tall as the first ones made. That means it is time to add more wax. I like to keep small pieces and flat wax flakes close at hand for these occasions.

TIP: Unlike big blocks of wax, small pieces and flat wax flakes melt quickly and keep the process of making candles running smoothly.

(continued next page)

(Height of desired finished candle + 2 inches [5 cm] loss on the bottom + 1 inch [2.5 cm] loss on top + ½ inch [1 cm] for wiggle room) x 2

So, if I wanted 8-inch (20 cm) finished length candles, I would cut my wick approximately 23 inches (58.5 cm) long.

In the above formula, I allowed a bit extra to tie a weight, such as a washer or nut, to the bottom of both wicks. This helps keep the wick straight through subsequent dippings. The extra weight will help to make a beautiful, straight candle.

WAX TEMPERATURE AND DIPPING METHODS

Temperature of the wax is very important. If the wax is too hot, the candles don't build up properly or build up more at the top than at the bottom. If the wax is too cool, the candles will look bumpy and have an exaggerated cone shape. I keep the wax in my dipping tank (wax tank) at about 170°F to 175°F (77°C to 79°C) and this temperature range seems to work well for me. How the candles look will depend on wax composition, temperature of the wax, temperature of the room, and individual style. It is best to play around with different dipping rhythms. Try dipping slow on the way down and pulling it back out fast. Then, try the opposite. Leave it at the bottom for an extra second or two. See how the shape of the candle changes with a couple dips. Experimentation is the key.

CHOOSE AND PRIME THE WICK

Dipped taper candles are a bit more forgiving with wick size than pillar candles. I have used a variety of wicks and almost all of them have burned well. I generally use a #2/0 or #1/0 square braid wicking for my dipped taper candles.

I prime all my dipped taper wicks by dropping a good quantity of wick into hot wax and letting it absorb completely before pulling it back out again. I look at the bubbles, and once they stop, I wait a bit longer and then remove the wick. As it begins to cool, I untangle the mass and start to straighten it out. If I am dipping pairs, I use the following general formula to cut my wick into useable lengths:

PREPARE A PLACE TO HANG DIPPED CANDLES

Only one more thing is needed before the dipping can begin—a place to hang in-process candles while they cool. I like to use paint sticks. They have a width that allows the candles to hang far enough apart so they don't to bump into each other, and the sticks are long enough that they can easily be placed across a box or between two chair backs, allowing the candles to hang freely.

HAND DIPPED TAPER CANDLES

TIP: Keep in mind that wax contracts as it cools. That means a freshly dipped candle that measures 3/4 inch (2 cm) will probably be thinner when it is completely cooled. Take that into account when making the candles.

1. When ready to begin, attach a nut or washer to each end of the primed wicks. Then, bend the wicks in half so that both sides are the same length.

2. Make sure that the wick is as straight as possible. Now, holding the wick at the bend, dip the wick into the wax. There is no need to do this quickly—I take 3 to 4 seconds to complete one dipping: 1 second down, 1 to 2 seconds at the bottom, and 1 second up.

3. Before moving on to the subsequent dipping, I wait until the wax changes color slightly, which takes 3 to 4 minutes. If I am doing multiples, I hang one up and move on to the others. By the time I complete one round of dipping all the wicks, the first one will be ready for round two.

TIP: Save all the trimmings. The wax can be melted off and the weights reused. Beeswax is very forgiving. If a candle doesn't turn out, put it back into the dipping vat. Once the wax has melted off the wicking, retrieve the wicking and start over.

4. Continue dipping until the candles have reached about 75 percent of the desired thickness. Wait for the wax to cool a bit longer, maybe 10 to 15 minutes, and then take a very sharp knife or scissors and cut the nut or washer off the end of the wick.

5. Complete a couple more dips to finish the candle and hang up to cool completely. If the dipping vat is on the stove, make the last dip a slightly hotter one (around 180°F [82°C]). The candle will be smoother and shinier, enhancing the overall look.

BIRTHDAY CANDLES

Birthday candles can be made a couple different ways. In essence, they are really thin taper candles, so they can be made the same way, stopping when the approximate diameter reaches 1/4 inch (6 mm). However, I prefer to make them to look more like the traditional birthday candles, which are squared off at the top and bottom. To achieve this, the candles are first dipped to a longer length and then trimmed to the desired length. Here is my technique.

HAND DIPPED BIRTHDAY CANDLES

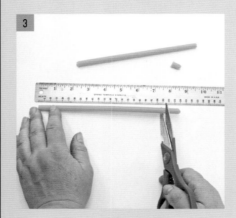

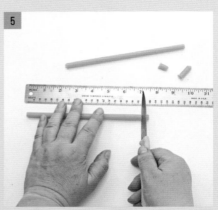

1. I begin by using a slightly thinner wicking: #2/0 or #3/0. I cut and prime the wick and stretch it out as I did for the tapers. I don't bother with a weight because I am only dipping them 6 to 8 times total.

2. Dip the candles 5 to 6 times and measure the diameter. If they are close to the thickness desired, dip one more time for good measure and let them cool until they just begin to turn from the pastel yellow to the more gold color, but are still a bit warm to the touch.

3. Use a scissors to cut the wick just above and below the desired length of the candle.

4. To make the 7-inch (18 cm) candles shown above, I cut my candles into 7 1/2-inch (19 cm) sections with a sharp pair of scissors. Then, I roll the cut sections between two sheets of wax paper on the table top to straighten and smooth the imperfections. The candles also need to cool a bit more before continuing to the final trimming.

5. Measure 7 inches (18 cm) from one end of the candle and with a sharp knife, cut into the beeswax while rolling it to cut the wax, but not the wick. Pull the little chunk of wax off the wick and the candle is completed. Repeat the process until the desired number of candles have been made.

MATERIALS

#2/0 or #3/0 wicking, cut to desired
length

Beeswax

Container for melting wax

Scissors

Sharp paring knife

Wax paper

Ruler

TIP: Keep in mind that birthday
candles can be made in a variety of
lengths depending on presentation.
Short candles can be used to blanket
a sheet cake while taller ones can
provide some elegance.

MOLDED CANDLES

Candles have been molded since the fifteenth century. Candle molds were originally made from metal, such as pewter, tin, and later aluminum. A recent addition, silicone rubber, has significantly expanded the number of possible candle shapes. I have candle molds made from a variety of materials. In this section, I will go into making each candle using my favorite type of mold.

TAPER CANDLES

MATERIALS

Square braid wicking, #1/0 or #2/0

Beeswax (enough to make the desired number of candles—see sidebar on page 56 for approximate amount needed for various taper lengths)

Silicone taper mold(s)

Rack to hold molds upright

Silicone mold release spray (optional)

Bobby pins

Beeswax pouring pitcher

Noncontact infrared thermometer

Wooden stir stick

As far as candles go, molded tapers are by far my favorite! Over the years, I have made thousands of taper candles, easily and relatively effortlessly, using my silicone taper molds. Although metal molds were available, I made the decision to purchase the silicone ones instead because I had heard stories of candles not always releasing from the molds easily.

One of the reasons I like molding my taper candles instead of dipping them is that 100 percent of the melted wax can be used for making the taper candles. With dipped candles, I always need to have that reservoir of wax. Molding also allows me to make candles in different colors, tinting only the wax needed for a specific project.

1. Set up the molds by pulling a length of wicking through the bottom of the mold. I like to leave the wicking really long because the mold automatically re-wicks by pulling up the next candle length of wicking as the hardened taper candle is pulled up from the mold. The silicone mold holds the wicking tightly enough on the bottom, so there's no need to apply any kind of mold sealer. To hold the wick taut at the top of the mold, I like to use traditional bobby pins.

2. Place the molds in something that will hold them upright without allowing them to bend. Once in place, I give each mold a quick spritz of silicone mold release spray. I have found that the mold release spray is more important with colored tapers than the uncolored natural tapers, as the colorant can sometimes stick.

3. Now, it's time to pour the candles. For silicone candle molds, I find that the wax temperature needs to be quite a bit hotter than for other candles. I generally aim for temperatures around 190°F (88°C). Give the wax a final stir with the stir stick and pour the wax down the middle of the mold. Try to keep the wax from running down the sides of the inside of the mold, as those drips might cool more quickly and mar the finish of the candle.

4. Allow the candles to cool. I usually wait about 30 minutes before trying to unmold. As they cool, the wax will shrink and create a divot at the top. Top off the candles with additional wax as needed. Once that wax cools, unmold the candles and pull up on the wick to remove the candle. The wicking for the next candle will automatically pull up from the additional wicking below the mold.

5. If the candles are still slightly warm, I like to pull them out of the molds and hang them by the wick off the side of the mold until they have cooled completely. Move the bobby pins from the bottom of the cooled candle to the top of the mold and cut the wick to the desired length.

6. To finish the candles, trim the wick as close to the bottom of the candle as possible. Smooth and straighten the bottom by heating on an electric skillet or do what I do and use a candle fluter.

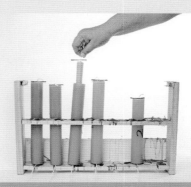

TIP: In order to create a beautiful candle with a smooth surface and symmetrical shape, it is best to let the candle cool down really slowly in the mold. One way to do this is to start with a warm mold. While this is not as important with smaller diameter candles, it helps to keep the candles from cooling too quickly, regardless of the candle size. I use felt cloth to help insulate the molds and a heat gun to warm them before pouring in the wax. Insulating the molds is optional, but it works well for me in my unheated basement.

CANDLE FLUTER

Because my candle molds vary slightly depending on size and shape, I use a metal fluter to give the bottoms a consistent diameter. The fluter is corregated so that the candle base will fit a range of holders. To flute candles, simply place the fluter in an electric skillet, heating the skillet to about 200°F (93°C). Once the fluter is hot enough to melt the wax, grasp the candle with both hands to keep it vertical and slowly push the candle down into the fluter. When the candle has hit the bottom of the fluter, pull it back up and allow the wax to drip off and cool before setting the candle aside. The excess wax will flow out of the hole at the bottom of the fluter, giving the base of the candle a nice, uniform look.

USING SILICONE MOLDS

Silicone taper molds are flexible. This is what allows them to release the candles with ease, but it also means that they need a bit of support to stay straight and remain upright. Commercial racks are available, but not required. For many years, I used a cardboard box that had holes a bit larger than the diameter of the candle mold to hold the molds. I now use a rack I built that resembles a test tube rack. It holds the molds in 2 to 3 places and suspends them above the tabletop by about 1 to 2 inches (2.5 to 5 cm) to allow the wicking to feed up through the mold continuously.

TEA LIGHTS

MATERIALS

Beeswax (enough to fill the desired quantity of tea light cups)

Tea light cups

ECO 1 1" Wick assemblies, or equivalent (See page 51.)

Hot glue gun

Wax melter

Noncontact infrared thermometer

Beeswax pouring pitcher

Honestly, before I started making beeswax tea lights, I really didn't see the need to make them. At the time, I considered tea lights as purely utilitarian candles that were sold in bags of 100 for a couple dollars. Why would anyone buy tea lights made from beeswax? A wholesale customer insisted, so I made a trial batch. I used the clear polycarbonate tea light cups and the finished effect was magical. What surprised me even more was how long they burned. They burned brightly for 2 to 3 hours or more!

My technique for making tea lights is a bit more involved that other methods I have seen, but because I sell these, I want to make sure that there are no surprises when the end user burns them.

TIP: Cover the work surface with craft paper, newspaper, or butcher paper. This will help with cleanup if some wax should accidentally spill.

1. To prepare the tea light cups, use the hot glue gun to adhere the wick assemblies to the bottom of the tea light cup. Take care to make sure they are centered.

2. Arrange the cups up on the edge of the table. Make sure the wicks are relatively vertical. Slight adjustments can be made after the wax has been poured.

3. Check the temperature of the wax before pouring. With tea lights, it's not as critical to get exactly the right temperature as when making other candles, but if the wax is too hot, it may overflow the cups more easily. If the wax is too cool, the surface may develop bubbles that will mar the top of the candle. I like to pour the first candle when the wax is around 170°F (77°C). The wax will cool as I continue to pour, so I stop and refill when it seems to be setting up a bit in my pouring pitcher. After pouring a couple tea lights, I nudge the wicks above the wax to make sure they are vertical and in the center and then proceed to pour wax into the next couple tea light cups.

4. Allow the candles to cool completely. Set aside for 1 day to cure before using the tea lights.

VOTIVE CANDLES

Together with tea lights, votive candles are the workhorses of the candle world. Votives are not intended to be burned by themselves, but should be set in a votive holder before lighting. This means that the candle should fit the holder reasonably well and the candle should be wicked so that the wax pool extends all the way to the edge. Most votives burn at least 6 to 8 hours.

1. Melt the beeswax in a double boiler or a wax melter. Do not melt beeswax directly on the stove, without the water bath. When I first started making candles, I used a clean coffee can set inside of a pot of water to melt my wax.

2. While waiting for the wax to melt, prepare the molds. If using metal molds, spray the inside with a mold release spray (silicone). Make sure to spray the bottom of the wick pins as well. Also make sure the pins are properly seated on the bottom of the votive mold.

TIP: Sometimes, simply pulling up on the wick isn't enough to release the candle from the molds. To facilitate removal, I like to push down on the top of the molds. Although this alone usually releases the candles, sometimes they need a bit of nudging at the bottom. For round molds, a slight twist will release it easily.

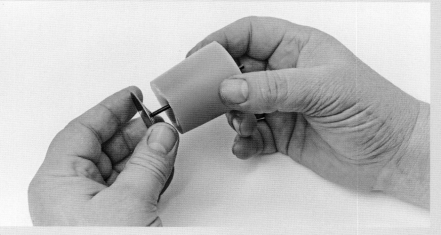

3. Once the wax is completely melted, use the thermometer to check the temperature of the melted wax. I like to pour my candles when the wax is in the 160°F to 165°F (70°C to 74°C) range. If it is hotter than that, let it cool a bit before pouring into the molds.

Optional: Wrap a layer of felt cloth around the molds to insulate them. I also use a heat gun to heat the mold right before I pour the hot wax in, to ensure that the wax doesn't harden too quickly where it touches the mold. Use the pouring pitcher to pour the hot wax into the molds.

4. Let the wax cool completely before trying to remove from the mold. Pull up on the wick pins to remove the candle and then tap the wick pin against a solid surface to dislodge it from the votive.

5. Thread a pretabbed wick up the hole left by the wick pin, trim the wick, and let it sit a day or so before doing a test burn.

PILLAR CANDLES

Pillar candles are the longest burning candles. Even shorter pillars measuring 3 to 4 inches (7.5 to 10 cm) have been known to burn sixty hours or more. They are well worth the investment in beeswax.

TRADITIONAL METHOD

HOW MUCH BEESWAX IS NEEDED?

The approximate amounts of beeswax needed to make various round candle sizes are as follows:

2" x 3" (5 cm x 7.5 cm) = 5 oz (140 g)

2" x 6" (5 cm x 15 cm) = 9 oz (255 g)

3" x 3" (7.5 cm x 7.5 cm) = 11 oz (310 g)

3" x 4" (7.5 cm x 10 cm) = 15 oz (425 g)

3" x 6" (7.5 cm x 15 cm) = 22 oz (625 g)

3" x 9" (7.5 cm x 23 cm) = 35 oz (995 g)

4" x 4" (10 cm x 10 cm) = 25 oz (710 g)

MATERIALS

Beeswax (enough to make the desired number of candles— see sidebar)

Pillar molds

Square braid wicking (try #3 for 2-inch [5 cm] diameter and #5 for 3-inch [7.5 cm] diameter molds)

Double boiler or wax melter

Silicone mold release spray (if using metal molds)

Scissors

Pliers (optional)

Wooden dowel, toothpick, or bobby pin

High temerature metal tape or wick putty

Beeswax pouring pitcher

Noncontact infrared thermometer

Heat gun

Felt cloth (enough to wrap around each of the molds once)

1. Melt the beeswax in a double boiler or a wax melter. Do not melt beeswax directly on the stove, without the water bath. Check the temperature of the wax occasionally to make sure it does not get too hot. Remember, the required pouring temperature is only 160°F to 165°F (70°C to 74°C).

2. While waiting for the wax to melt, prepare the molds. If using metal molds, spray the inside with a silicone mold release spray.

3. Most pillar candle molds come with a hole at the bottom of the mold for the wicking to pass through. I find it is easiest to first dip a part of my wicking in beeswax and let it harden and then trim the end at an angle to facilitate threading the wick through the hole and far enough into the mold to be able to retrieve it from the open end. A pair of pliers can help with this if the mold is a bit deeper than can easily be reached.

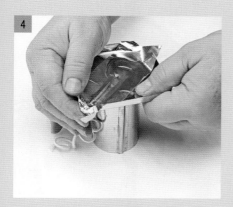

4. Pull the wick so that it extends about an inch (2.5 cm) or so above the candle mold. A wooden dowel, toothpick, or even a bobby pin can be used to hold the wick in place and to keep it centered. Just wrap the wicking around the holder and make sure it is tight inside the mold, yet not stretching the wick. The bottom of the mold will also need to be sealed where the wicking extends through the mold. There are a number of products that can be used for this. My favorite is high temperature metal tape that I buy in the heating section of the home improvement store. Another sealer that I sometimes use is something called *wick putty*, which is malleable dough that conforms to the bottom of the mold and seals the mold around the wick, preventing the beeswax from leaking out the bottom.

5. Wrap the molds with a layer of felt cloth and set them in a configuration that will best facilitate the pouring of wax into the candle molds quickly and efficiently. I like to do a dry run with my pouring pitcher to make sure I don't have to pour the hot wax from a point that is too high up because I have placed another candle in the way.

6. Once the molds are prepared, check on the wax. The target temperature should be in the range of 165°F to 170°F (74°C to 77°C). If it is hotter than that, let it cool a bit before pouring into the molds.

7. Use a heat gun to briefly warm the inside of all the pillar molds and immediately pour the wax into the waiting molds. Stop pouring when the wax is about a half inch (1 cm) from the top of the mold—do not fill the molds all the way to the top. Also, make sure there is enough wax in the pouring pitcher to fill the mold completely. The line between pours will be visible if the candle is poured in several stages.

8. Allow the candles to cool completely before trying to unmold. I like to do this the following day, especially when making a larger diameter candle.

9. To unmold, first take the tape or wick putty off the bottom of the mold. Then, lightly tap the side of the mold against a soft surface, rotating the mold while tapping. Once the candle is loosened from the mold, tug gently on the wick and remove the candle. If it isn't cooperating, don't force it. Put the whole works in the freezer and let it cool for an hour or so. Then try again. This trick should work. Trim the wick flush at the bottom and to a length of 1/4 inch (6 mm) at the top.

10. The bottom of the candle may be bumpy and uneven. To fix that, I like to use a dedicated electric skillet to melt

the bottom of the candle. Start with the temperature of the skillet on low and see if that will melt the wax on the candle. Electric skillets vary by manufacturer, so a bit of experimentation is necessary. If low doesn't melt the wax, increase the temperature until it starts to melt the wax relatively quickly. Place the candle in the skillet and spin it so that the high points in the candle bottom become more evident. Failure to do this will result in a leaning candle that will burn unevenly. Once the bottom is smoothed out and even, let it cool slightly and then clean up the edges by running fingers around the edge to get rid of errant bits of wax.

11. Now comes the hard part: the cure. Wait at least a day or two for the candles to cure before lighting them. That will allow enough time for the beeswax molecules to align and settle down. With pillar candles, a burn test is imperative to make sure that the right wick size is used. It could take five or more tries to get the wicking right.

QUICK COOL METHOD

MATERIALS

Beeswax (enough to make the desired number of candles—see sidebar on page 66)

Pillar molds

Square braid wicking (try #3 for 2-inch [5 cm] diameter and #5 for 3-inch [7.5 cm] diameter molds)

Double boiler or wax melter

Silicone mold release spray (if using metal molds)

Scissors

Pliers (optional)

Wooden dowel, toothpick, or bobby pin

High temerature metal tape or wick putty

Beeswax pouring pitcher

Noncontact infrared thermometer

Heat gun

Small stationary fan

I just recently discovered this method for making pillar candles, and it has changed my life. Rather than making sure that the candles cool as slowly as possible and monitoring them for cavities to be filled, the quick cool method relies on a fan to remove the heat from the surface of the candle molds. The candles cool on all sides rather than just the top. The wax shrinks just enough to prevent cavities from forming on the inside and the candles can be unmolded after only a couple hours. This is not the same thing as putting the candles in a cold environment to cool them quickly; instead, you simply remove the hot air that surrounds the candle.

This technique requires quite a bit of experimentation and knowledge of how wax responds as it cools, so I consider this an advanced technique. I recommend that novice pillar candle makers use the traditional method first to acquaint themselves with the candle making process.

Most of the steps involved with this process are the same as with the traditional method, but instead of insulating the molds, the fan is used. This technique requires quite a bit experimentation to get the setup right. The resulting candles are a bit thinner than the more traditionally made ones because the wax is allowed to shrink, but once the kinks have been worked out, this method produces an equally nice candle.

To set up the candle making area, I arrange my empty prepared molds in a line perpendicular to where my fan will be. The fan placement will vary depending on how powerful the fan is and how far from the molds it is. My fan is an inexpensive small stationary fan that I place approximately 4 feet (1.2 m) from the molds. I use the lowest setting and check the air flow with a thin strip of paper. The paper moves when it is in the range of the air flow of the fan. Once the fan is in the right location, I temporarily turn the fan off and then turn the fan back on after I fill my molds.

1. Melt the beeswax in a double boiler or a wax melter. Do not melt beeswax directly on the stove, without the water bath. Check the temperature of the wax occasionally to make sure it does not get too hot. Remember, the required pouring temperature is only 160°F to 165°F (70°C to 74°C).

2. While waiting for the wax to melt, prepare the molds. If using metal molds, spray the inside with a silicone mold release spray.

3. Most pillar candle molds come with a hole at the bottom of the mold for the wicking to pass through. I find it is easiest to first dip a part of my wicking in beeswax and let it harden and then trim the end at an angle to facilitate threading the wick through the hole and far enough into the mold to be able to retrieve it from

the open end. A pair of pliers can help with this if the mold is a bit deeper than can easily be reached.

4. Pull the wick so that it extends about an inch (2.5 cm) or so above the candle mold. A wooden dowel, toothpick, or even a bobby pin can be used to hold the wick in place and to keep it centered. Just wrap the wicking around the holder and make sure it is tight inside the mold, yet not stretching the wick. The bottom of the mold will also need to be sealed where the wicking extends through the mold. There are a number of products that can be used for this. My favorite is high temperature metal tape that I buy in the heating section of the home improvement store. Another sealer that I sometimes use is something called *wick putty*, which is malleable dough that conforms to the bottom of the mold and seals the mold around the wick, preventing the beeswax from leaking out the bottom.

TESTING

If this technique is not done properly, a large void can be created inside the candle. This may cause problems when the candle burns, by allowing melted wax to drain into the cavity and exposing a large amount of wick, creating a large flame.

To test candles to make sure there are no cavities inside, I suggest opening up the candle at the bottom with a hot knife and doing some forensic investigation. If there is a cavity inside the candle, the fan is not removing enough heat. To remedy this problem, either move the fan closer or turn it up.

5. Once the molds are prepared, check on the wax. The target temperature should be in the range of 165°F to 170°F (74°C to 77°C). If it is hotter than that, let it cool a bit before pouring into the molds. (See step 1 on page 66.)

6. Use a heat gun to briefly warm the inside of all the pillar molds and immediately pour the wax into the waiting molds. Stop pouring when the wax is about a half inch (1 cm) from the top of the mold—do not fill the molds all the way to the top. Also, make sure there is enough wax in the pouring pitcher to fill the mold completely. The line between pours will be visible if the candle is poured in several stages.

7. Make sure that the stationary fan is in the correct position and turn it on.

8. The wax on the top of the candle mold will start to solidify soon after the wax is poured. Once most of the top has skinned over, after about 5 to 10 minutes, I rotate the mold about a third of the way. After another 5 to 10 minutes, I rotate the candles another third of the way in the same direction. I keep checking and turning at regular 5 to 10 minute intervals until the molds feel cool to the touch.

9. At this point, the wax will be pulled away from the sides of the candle mold and the wax height will have shrunk down slightly from where it was when the wax was poured initially. It should now be safe to unmold the candle. I untape the bottom and carefully remove the candle. Take care to ensure that the wick stays put; the inside of the candle may still be quite warm, so the wick may pull through instead of staying inside the candle. Once the candle is removed from the mold, allow it to cool completely before finishing it using the same technique as with the traditional method.

DEALING WITH CAVITIES

As the wax cools, it will contract. How evenly it contracts will also depend on the ambient temperature of the room. Cooler temperatures will force the top of the candle to skin over and become a solid mass before the rest of the candle can contract sufficiently. The result is a cavity inside the candle. Bigger candles have greater mass and will take longer to contract, increasing the likelihood for the candle to develop a large cavity. This can be dangerous if left as is necause the wick can turn into a torch when it hits the open cavity.

To counteract the cavity, I make a relief hole or widen an existing hole with a nail or piece of toothpick. Then, I pour melted wax into the hole. It may not look like much of a hole, but a surprising amount of wax will likely be needed to fill it. Repeat if necessary before the wax cools completely.

CHAPTER 4
ALCHEMY FOR THE HOME

For items that are used around the house, beeswax reigns supreme. It shines, both figuratively and literally, as a furniture and wood polish or conditioner. It also transforms fabric into a reusable food protector. The sole outlier in this section is the Woodcutter Incense, which uses propolis as one of its star ingredients.

FURNITURE POLISH

Beeswax furniture polish is considered the ultimate in wood care, due in part to its soft, satin shine. The touch of carnauba wax results in a harder, shinier finish.

MATERIALS

Beeswax, yellow	1.5 oz	43 g	10%
Carnauba wax	0.4 oz	10 g	2.4%
Turpentine	13.1 oz	372 g	87.6%

Double boiler

Wooden stir stick

Wide-mouth jars with lids (shorter jars work better for scooping out polish)

Yield: Makes approximately 16 fluid ounces (475 ml)

1. Melt the waxes in a double boiler.

2. Remove from the heat and stir in the turpentine.

3. Once it is completely mixed, pour into wide mouth jars.

4. Apply the polish with a clean cloth and rub in small circles. Turn the cloth as it becomes dirty. Allow the polish to dry and then buff the same area with a clean cloth. If more than one coat is desired, wait two days between applications.

WHAT IS TURPENTINE?

Pure gum turpentine is a natural derivative of pine resin. It has a strong odor and care should be taken not to inhale the fumes. It is possible to buy low odor turpentine.

SALAD BOWL CONDITIONER

For food related utensils, such as salad bowls and wooden spoons, a more food-friendly concoction is required. I swear by mineral oil, which has an almost indefinite shelf life and won't go rancid. Recently, however, I have tried some of the other suggestions, such as walnut oil, and I have to say it is nice. With this recipe, either oil works fine.

1. In a double boiler, combine the beeswax and oil and heat until the beeswax is completely melted. Stir until thoroughly mixed.

2. Pour into small wide-mouth jar or tin.

3. To use, simply rub a small dollop into the salad bowl with hands or a clean cloth and wipe with a dry cloth.

MATERIALS

Beeswax, yellow	1.8 oz	50 g	25%
Mineral oil (or walnut oil)	5.3 oz	150 g	75%
Double boiler			
Wooden stir stick			
Small wide-mouth jar or tin			
Yield: Makes approximately 8 fluid ounces (235 ml)			

WOOD CONDITIONER CREAM

MATERIALS

Beeswax, yellow	5.3 oz	60 g	28.5%
Turpentine	2.1 oz	150 g	71.5%
Double boiler			
Wooden stir stick			
Wide-mouth jar with lid (shorter jars are easier to scoop from)			
Yield: Makes approximately 8 fluid ounces (235 ml)			

This recipe is perfect for nourishing the wood. It cleans and protects in one step.

1. Melt the wax in a double boiler.

2. When melted, remove from the heat and stir in the turpentine.

3. Mix well and pour into a wide mouth jar.

WAXED COTTON FOOD WRAPS

I have included two recipes for food wraps: one uses just beeswax, while the other utilizes a blend of beeswax, resin, and oil. The technique for waxing the fabric are the same. I personally prefer the beeswax-only recipe for sandwich wraps, but as a substitute for plastic cling film, I prefer the beeswax/resin blend.

CHOOSING THE RIGHT FABRIC

Part of the success of food wraps relies on making good choices when selecting the fabric. Look for highly absorbent natural fibers such as cotton and linen. Also, look at the tightness of the weave. A thin, gauzy material will not keep food fresh, regardless of how much wax is used. Conversely, a heavy canvas, once waxed, may be too difficult to bend around the food. For sandwich wraps, I like to use a medium-weight cotton twill. For cheese wraps, I prefer a lightweight cotton.

HOW TO APPLY THE WAX

Everyone has a preferred technique for applying the wax onto the fabric, whether it's dipping the fabric into the wax, applying the wax with a brush, or sprinkling grated wax over the surface of the fabric and heating it. Personally, I prefer to use an electric skillet to melt the wax and then submerge the fabric in the wax, so that the fibers fully soak up the wax. The actual amount of wax or wax/resin mixture needed depends on the size of the electric skillet. One pound (455 g) of beeswax should be plenty for a bunch of wraps.

MATERIALS

Beeswax-Only Food Wraps:

Fabric, cut to desired size— see sidebars

Beeswax, approximately 1 pound (455 g)

Double boiler or small electric skillet

Wooden stir stick

Heat gun

Sewing machine and thread (optional—see sidebar)

Beeswax/Resin Blend Food Wraps:

Fabric, cut to desired size— see sidebars

1 pound (455 g) beeswax

3 ounces (85 g) pine gum rosin

1/2 ounce (15 g) jojoba oil

Double boiler or small electric skillet

Wooden stir stick

Heat gun

Sewing machine and thread (optional—see sidebar)

Yield: Variable, depending on the size of your fabric and electric skillet or double boiler.

2

SIZES TO TRY

Small items = 8" x 8" (20 cm x 20 cm)

Cheese size = 10" x 10" (25.5 cm x 25.5 cm)

Sandwich size = 12" x 12" (20 cm x 30 cm)

Baguette = 15" x 26" (38 cm x 66 cm)

Bread = 18" x 24" (46 cm x 60 cm)

1. Melt all the ingredients in a double boiler or small electric skillet over low heat. Stir to ensure ingredients are thoroughly combined. Take care not to overheat. The mixture is now ready for fabric waxing.

2. Holding on to one corner, dip part of the fabric into the wax. Use a wooden stir stick to help submerge the fabric if needed, continuing to hold one corner up out of the wax. Pull the fabric out of the wax and allow the wax to drip off and the fabric to cool slightly.

3. Now, dip the unwaxed portion of the fabric into the wax. Lift it out and allow it to drip and cool slightly.

4. Before the fabric cools completely, stretch and lay the fabric flat on a work surface. At this point, there is probably a surplus of wax on the fabric. Go over the fabric with a heat gun and some paper toweling to absorb some of the excess wax. While it's still warm, pull the fabric in all directions to straighten the grain of the fabric. Allow to cool completely.

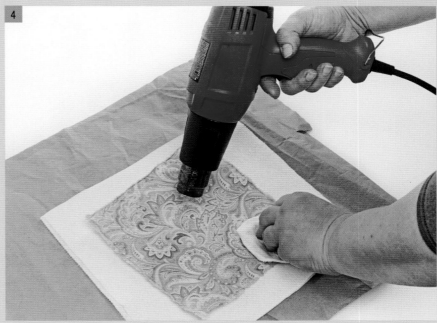

4

To finish the cut edges, I like to use my sewing machine to add a zigzag stitch around the perimeter. Although not really necessary to keep fraying in check, I find it gives the wraps a more finished appearance.

WOODCUTTER INCENSE

Growing up, my only real exposure to incense was the German woodcutter incense smoker my family used over the holidays. It used incense cones, which gave off a lovely, woodsy aroma when burned. Later in college, I had friends who occasionally burned incense sticks. They smelled horrible and gave me a pounding headache. At the time, I thought perhaps my memory of the incense from my childhood was wrong, and that all incense smelled like the sticks. What I know now is that there is a huge difference between well-made incense that rely on natural resins, woods, and other botanical ingredients and the cheap incense for sale at the convenience store.

A few years ago, I learned that honey and propolis are both really good ingredients for incense. I decided to try my hand at making them. I'm not sure if I achieved what I remembered from childhood, but I am happy with my results. I also revel in the collecting of local materials, such as pine needles and rose petals, for my incense and the grinding and molding of cones that reflect who I am today and the world in which I live. The following recipe is the one I call my Woodcutter Incense. Enjoy!

MATERIALS

1/2 teaspoon propolis ground

1/4 teaspoon dragon's blood powder

1 tablespoon (approx. 15 g) red sandalwood powder

1 teaspoon cherry bark powder

1/2 teaspoon makko powder

1/4 teaspoon orris root powder

1/4 teaspoon orange peel powder

1/2 teaspoon juniper berries, crushed

1/2 teaspoon ceremonial sage, powdered

1/2 teaspoon pine needles, powdered

1 teaspoon rose petals, powdered

Honey, as needed

Water, as needed

Mortar and pestle

Wax paper

Yield: Makes approximately 12 cones

1. Make sure all dry ingredients are finely ground. For wood and tougher items such as pine needles, chopping them first in a spice grinder helps. Once all the ingredients are at least a rough powder, grind them together to create a relatively homogenous powder.

2. Pour most of the powder onto the wax paper. Reserve just a little in case too much liquid is added. Create a well in the middle of the powder and add a couple drops of honey, but not too much or the incense will smell like burnt sugar. The honey is usually not enough to make a dough, so I add water in very small amounts until a malleable dough is formed. Add in the remaining powder and a drop or two of water and the dough should be ready for forming.

3. The cones are best when made proportionately tall and narrow. This shape allows them to dry out quickly and thoroughly and is a good shape for burning. Each cone uses approximately 1/4 teaspoon of dough.

4. Stand the incense cones upright on the wax paper and place them in an out-of-the-way place to dry out for at least 2 days. Once dry, store them in an airtight container until ready to use.

SCENTED BEESWAX MELTS

Melts or tarts are chunks of scented wax that melt easily over a tea light candle (my preference) or a specially-made electric warmer. The idea is that the heat from the candle melts the wax, which in turn scents the air. I prefer scenting wax in this way rather than adding scent to candles directly. One reason is that I prefer to use essential oils rather than fragrance oils. Essential oils aren't really made to be burned. At best, the scent can change or dissipate. At worst, the essential oil can separate from the wax and cause the flame to flare up. With beeswax melts, that isn't a problem. The warmth of the tea light flame is just enough to allow the wax to release the wonderful scent.

In order to produce a melt that is hard enough to keep its shape when cool yet soft enough to melt readily over a small flame, the beeswax needs to be mixed with an oil. For this, I like to use palm kernel oil. It is solid at room temperature, but melts easily at 75°F (24°C).

I like to mold these into individual deli cups, but they can also be molded in small silicone molds.

MATERIALS

Palm kernel oil	3 oz	85 g	42.86%
Beeswax, yellow	3 oz	85 g	42.86%
Essential oil blend	1 oz	28 g	14.29%
Double boiler or microwave and container			
Wooden stir stick			
Silicone mold or 1 ounce (28 ml) deli cups			
Yield: Makes approximately 8 melts (1 ounce, or 28 ml, each)			

1. Heat the palm kernel oil in a double boiler or in the microwave until completely melted.

2. Add the beeswax and continue to heat until it is also melted. Remove from heat and allow to cool to about 125°F (52°C). It should still be liquid, but not hot. Add the scent and stir well.

3. Pour the mixture into silicone molds or deli cups.

SEALING WAX

MATERIALS

2 ounces (55 g) beeswax

2 ounces (55 g) damar resin

2 ounces (55 g) shellac

Marble dust or plaster powder, as needed

Earth or other coloring pigment, as desired

Double boiler

Wooden stir stick

Mold or aluminum foil folded into a v-shape

Yield: Makes approximately 4 to 5 sticks

I always thought of sealing wax as something archaic or something that might be used only for wedding invitations. Then, I had an idea: instead of using plastic shrink bands on our honey jars, how about using sealing wax? It took quite a few trials to get it right, but I finally achieved the effect I was after. It is a fun way to use a traditional technique. To mold them, look for silicone molds used to form ice sticks for water bottles.

1. Weigh the beeswax, resin, and shellac and add to the top of a double boiler. Warm over low heat until the resin has melted. They are incompatible materials, so there will be separation between the shellac and the other ingredients. This is okay.

2. Begin to sprinkle in the marble powder, going slowly and stirring to avoid creating clumps. The marble powder facilitates the bonding of all the ingredients.

3. When the layers have started to homogenize, begin sprinkling in coloring pigment of your choice. How much color depends on the color intensity of the pigment. Each pigment will require a different amount. Color can be tested by dabbing some hot wax on piece of scrap paper. The hardened seal will be slightly less vivid than the molten wax.

4. Pour the wax into molds or aluminum foil.

CHAPTER 5
ALCHEMY IN THE STUDIO

Beeswax can be a lot of fun to use in a variety of different art projects for all ages. The projects included in this section range from making your own crayons; to encaustic painting—creating works of art using hot wax and pigments; to batik, the ancient art of using wax and dye to create designs on fabric. For the encaustic and batik projects, use the instructions provided here as a springboard to inspire your own works of art.

BEESWAX CRAYONS

MATERIALS

Beeswax, yellow	21.6 g	36%
Jojoba oil	4.2 g	7%
Bentonite clay	8.4 g	14%
Earth pigment powder	25.8 g	43%
Double boiler or electric skillet		
Wooden stir stick		
Molds		
Silicone mold release		
Yield: Approximately 5 crayons		

Crayons may seem to be easy things to make—a little wax, a little pigment. In actuality, they need to be finely tuned "tools." If the crayon is too soft, it leaves clumps of crayon on the paper. If it is too hard, it barely leaves a mark. The quality of the mark also matters—a lot of crayons don't deposit color evenly on the paper.

There is also the issue of what to use for pigment. A lot of the pigments available are great for adding to oil paint, which it will rarely get on the skin and is unlikely to be ingested. Crayons, on the other hand, are often used by children, who have been known to eat them and color on things not intended to be colored. Most of the pigment is technically bound into the base wax, so I don't believe skin contact is a problem, but crayons are certainly not meant for eating. Most of the food-safe colorants are water soluble or water based, which makes using them in crayons difficult, if not impossible, because wax doesn't play well with water. For these crayons, I use Earth pigments, which are safe for skin, but should not be ingested. For this reason, these crayons are not intended for use by really young children.

My crayon recipe takes inspiration from the make-up world. Lip and eyeliner pencils are safe on sensitive facial skin and produce a nice, smooth line and deposit color without applying too much pressure.

1. Melt the beeswax in a double boiler or electric skillet. Add the jojoba oil and bentonite clay. Mix with a small wooden stir stick until thoroughly combined.

2. Add a small amount of pigment and stir, gradually increasing the amount until desired shade is reached. The amount shown in the recipe is just a rough guideline for color. Don't judge the color based on how the wax looks in its molten state. There is a difference between actual color of a crayon and how much color the crayon deposits on the paper. Allow a small dollop of wax to harden and try drawing on a sheet of paper to test color strength.

3. Pour the wax into molds that have been sprayed with mold release. Allow to cool completely. Test a crayon for color intensity and ease of application. If the crayon recipe needs to be adjusted, they can easily be remelted and tweaked as needed. If they are too hard, add more oil. If they are too soft, add more beeswax.

ENCAUSTIC

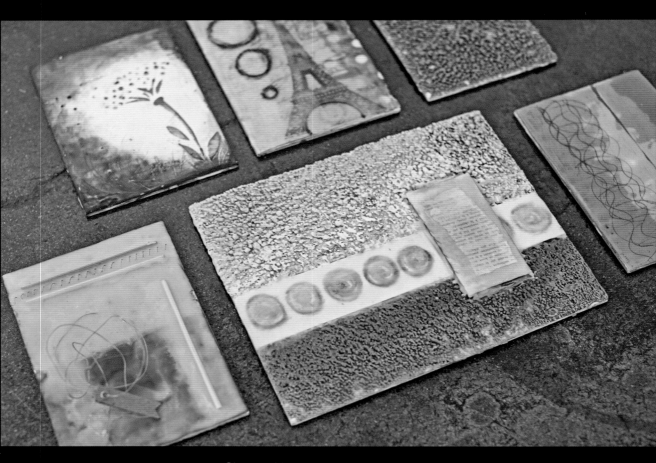

Since writing my last book, *Beeswax Alchemy*, I have had many opportunities to play with encaustic. In the process, I have learned that it is an incredibly malleable medium that is, at the same time, both extremely versatile and in some ways, frustrating. I have found that simply having a vague sketch of an idea is better than having the entire piece planned in detail. The result rarely looks as I expect it to, but it usually does turn out better than I had hoped. The wax teaches patience and acceptance.

It also helps to know that if the plan doesn't work out, a heat gun and a scraper can take the artwork back to square one.

Although I have included a project for encaustic in this book, my goal is not to have users replicate my project step by step. Instead, my intention is to illustrate how the different techniques come together to create the artist's vision. The techniques I illustrate are just a handful of useful ways to manipulate wax. Almost anything goes,

TOOLS FOR ENCAUSTIC

Encaustic work requires a few basic tools. Here is what you will need to get started.

FUSING TOOLS

Each layer of wax needs to be fused to the layer beneath it. The best tools for fusing are a heat gun or a propane torch, but irons created for this purpose can be used as well. My personal choice is a heat gun, but I have a torch that comes in handy for small areas that require a bit more precision.

HOT PALETTE

Encaustic medium and encaustic paints need to be melted before they can be applied. The easiest way to do this is to place them on a heated palette. There are specially designed heated palettes for encaustic work that are completely flat and give the user complete control of the temperature. These palettes are definitely nice, but a cheaper option is to use an electric griddle. I chose one that had a light surface, so I could mix some on-the-spot color directly on the griddle surface.

BRUSHES

The brushes for applying encaustic medium can be any shape or size, but they need to be natural. My personal favorites are Japanese Hake brushes. They are relatively inexpensive and seem to hold wax really well. They also come in a variety of widths.

OTHER TOOLS

There are no limits to the number of fun tools that can be used to manipulate the wax. Check the utensil drawer in the kitchen and the toolbox in the garage for novel ideas. Some I have used include the following:

Pottery shaping tools: These can be used to gouge or remove warm wax.

Spoons: The rounded back side of a spoon works well to help burnish paper to the wax surface.

Wire brushes: Clean wire brushes can be used to scratch the surface of the wax, creating texture.

Wood burning tools: These tools can be used to melt small holes or larger surfaces of wax, without disturbing neighboring wax.

ENCAUSTIC SURFACES

Generally, encaustic art pieces need to be created on rigid surfaces. If there is any flexibility in the substrate, the layers of wax are at risk of cracking and chipping off. Ideal encaustic surfaces also need to be absorbent. If the wax does not soak into the surface, the artwork has nothing to adhere to.

There are boards specially designed for encaustic work that come already primed and ready for wax. Encaustic panels are often cradled boards, which means there is a wood surface attached to a wood frame ($\frac{3}{4}$-inch [2 cm] and 2-inch [5 cm] deep frames are common). The frame allows the panels to be hung directly on the wall without the need for framing and gives the artwork a more modern feel. These are nice, but not really required.

I like to use inexpensive wood panels from the craft store, especially for playing. They are nice and smooth and rigid. Perfect!

PREPARING THE SURFACE

Before starting my encaustic projects, I usually prepare the surface by laying down at least 3 to 4 layers of encaustic medium, either clear or tinted white. There are specially made encaustic primers out there, but they are usually just paint made without latex. An inexpensive option is classic tempera paint. Applying a couple layers of tempera paint before applying wax creates a nice white background.

ENCAUSTIC MEDIUM

Encaustic medium is refined beeswax combined with damar resin. The best beeswax to use for this is what I refer to as *white* beeswax. The beeswax has been specially filtered to remove all the yellow color. Damar resin is a resin that comes from the Canarium strictum tree, which grows in India and East Asia. Damar resin is added to the beeswax to raise the melting temperature of the medium, which protects the artwork from damage. The resin also adds luster, luminescence, and translucency to the artwork once it is fully cured.

Although encaustic medium and encaustic paint can be purchased in a ready-to-use form, being a DIY kind of gal, I prefer to make my own. It takes a while for the beeswax and resin to melt, so I like to make a big batch a day or so before I actually plan to paint.

There is artistry in the making of the medium. The desired finished effect will determine the exact ratios of beeswax to resin. If too much resin is used, the medium will become brittle and could flake off. If too little resin is used, the piece will remain a bit soft, leaving it prone to dust accumulation. I like the proportion of 9 parts beeswax to 2 parts damar resin. Others prefer a ratio of 10:1. Some use no damar resin at all.

WHY WHITE BEESWAX?

Although it is possible to use traditional yellow cappings wax, I strongly urge the use of white beeswax. I started off my journey with encaustics using what I had on hand—yellow beeswax. I added my pigments to that and thought I could make the yellow undertones a feature and work with it, but what I ended up with instead was muddy colors and very childish looking artwork. It was a waste of wax and pigment! I now use only white, highly refined beeswax and have had a much easier time creating works that I am proud of.

ENCAUSTIC MEDIUM MASTER BATCH

MATERIALS

3.5 ounces (100 g) damar resin

16 to 18 ounces (455 to 510 g) white beeswax

Electric skillet with temperature control

Wooden stir stick

Bread loaf pan with nonstick coating

Large knife

1. Melt the resin in the electric skillet. Once the resin is melted, add the beeswax and stir. Keep stirring occasionally until all the wax is melted and all the resin is incorporated. This will take a while, so be patient. There will be some impurities in the resin that will not melt. Ignore them for now.

2. Pour the wax and resin mixture into the loaf pan and allow the wax mixture to solidify. Once it is hard enough to remove from the loaf pan, but still slightly warm to the touch, invert it on a protected surface. All the impurities that were in the resin will have fallen to the bottom and are now visible. With the large knife, either cut off the bottom part with all the impurities or just scrape the bottom with the knife until all the impurities are gone and set aside. I like to keep the wax scrapings and add them to the next batch of encaustic medium that I make. That way, nothing goes to waste.

3. With the knife still in hand, I cut the clean portion into smaller cubes that are easier to handle if used as is or are the perfect size to toss into a tin and mix with pigment.

TIP: Damar resin is easily purchased at art supply stores, either locally or online.

ENCAUSTIC PAINT

There are many options for adding color to encaustic works and creating an encaustic paint is just one way. The best method varies depending on what color is needed and how much saturation of color is desired. The color can come from powdered pigments or from oil paints, if those are already on hand. To use oil paint, simply squeeze out a line of paint onto a sheet of paper towel and allow the towel to absorb some of the oil before mixing it with the encaustic medium. My preference is to use French mineral pigments, which have some transparency and are available in a nice range of colors that aren't too primary, but any kind of powdered artists' pigment will work.

If I am making saturated, super-strength colors, I like to work with them in 4 ounce (120 ml) flat-bottomed tins, either blending them with encaustic medium directly on the electric griddle or adding the concentrated color to a clean mini loaf pan and diluting it as desired.

Because I make my own medium and I have already cut it into useable pieces, I just toss a couple chunks into a mini loaf pan and put the pan on the hot griddle. Then, I add a touch of pigment and stir with my brush until it is fully mixed. To test the color density, I keep a piece of watercolor paper on hand. It is easier to add more pigment, rather than increasing the size of the color batch by adding additional encaustic medium. The loaf pans are ideal for this use, as the high sides support the brushes while the pan heats on the griddle.

Also, pay close attention to the temperature of the griddle and the wax. The ideal temperature for encaustic work is 200°F (93°C). If the temperature rises above 220°F (104°C), the wax can start to smoke and degrade.

ENCAUSTIC TECHNIQUES

There are probably as many encaustic techniques as there are encaustic artists. Every artist develops a repertoire of tools and techniques that work with his or her individual style. My goal here is to highlight a few techniques that are easy to do and can result in an interesting art piece, even for complete novices.

Here is both smooth and dry-brush texture.

TECHNIQUE 1: SMOOTH OR TEXTURED

Generally, there are two surfaces in encaustic: smooth and textured. For a smooth surface, start by warming the board with the heat gun. Use a paintbrush to apply a single coat of encaustic medium or encaustic color. Fuse the wax with a heat gun by warming the wax enough to make it shiny, but not so much that the wax begins to move it around. Apply another coat of wax and again fuse the surface. The surface should begin to look smooth, and if there is color, the color should look uniform. It may take 3 to 5 coats to achieve a really smooth surface.

To texture the surface, there are a lot of options. An easy one that yields interesting results is the dry-brush technique. This technique can be done on a warm surface or a cold surface, with different results. On a warm surface, the results will be more subtle, while a cold surface will yield a more pronounced effect. Apply color with a coarse brush, either by dabbing or using brush strokes. Lightly fuse with a heat gun and continue adding layers of dry brush paint, fusing lightly between coats, until the desired effect has been achieved. This technique often works better using a variety of colors to add depth to the texture.

TECHNIQUE 2: SCRAPING

Scraping is a technique that involves the removal of one or several layers of wax to reveal the layers below. To do this, apply multiple layers of wax in various colors, fusing after each layer. The layers can be a bit uneven, and that uneven texture will reveal an interesting end result. Once the board has cooled completely, use a single-edge razor blade or other scraping tool to begin revealing the layer below. Don't try to remove too much with each scrape. Vary the direction of the blade and the stroke. Continue scraping until the desired effect has been achieved.

TECHNIQUE 3: INCISING

Incising is a really easy way to add interest to an encaustic piece. Incising involves scratching, cutting, or melting the wax with tools. I have a set of clay tools with a variety of shapes that I use for incising. Coarse sandpaper or a wire brush can also be used to scratch lines into the surface. Once the marks have been made, there are a couple ways to highlight the lines.

Small, thin lines scratched into the surface can be highlighted by rubbing some darker oil paint sticks into the lines. Use vinyl gloves to keep your hands clean and rub the color into the scratches. To remove the excess color from outside the incised lines, apply a little vegetable oil to a sheet of paper towel and rub away the color. Once it has been cleaned as much as desired, fuse with a heat gun to seal it into the surface.

Pottery carving tools can also be used to create thick lines in the artwork. The lines are gouged into the wax and then filled with a contrasting color wax. The waxed lines can then be scraped back with a single-edge razor blade, until the inscribed line is even with the surrounding surface.

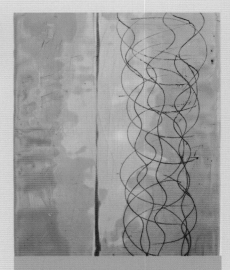

Two techniques are shown here. The scraping technique is used on the wax down the middle, and the incising technique is used two ways: a deep line on the left and scratching on the right.

Here are two techniques: stenciling and adding color using PanPastel.

TECHNIQUE 4: STENCILING AND MASKING

Stencils and tape are two ways to contain an encaustic wax application. Stencils can be used to add shape and texture to a piece. Simply place the stencil on the artwork, burnishing it to help keep it in place and to make sure the entire stencil is in contact with the art piece. Carefully apply a layer of wax. Remove the stencil before the wax hardens completely to keep the pattern from lifting off with the stencil. Stencils can also be used with nonwax colors, such as PanPastels, to apply a pattern without adding more layers of wax. With all stencil work, use a heat gun to fuse into place. Smaller stencil patterns can also be used as a starting point for creating additional texture with the dry-brush technique.

Tape can be used similarly, as a way to confine wax to a specific area or to mask an area that needs to stay free of wax. When using tape, it is important to make sure the tape adheres well to the surface of the work, especially along the edges. If there is texture, it will be difficult to get a good seal. Remove the tape after the wax has been fused, but before it cools off completely.

TECHNIQUE 5: ADDING COLOR

Sometimes, it is desirable to add color to the piece without adding more wax. One way to accomplish this is to use PanPastels. They are soft pastel colors compressed into a shallow cakelike pan. They can be applied using a soft brush, but I prefer a foam sponge or applicator for encaustic work. I add the color into the desired area and rub it in with my finger, burnishing the color in place. Fuse the area with a heat gun or torch to seal the color in place. Additional color can be layered on top to great effect. If something doesn't look right, it can be wiped away with a touch of rubbing alcohol and a paper towel. Nothing is permanent until it is fused into the wax layer.

Various weights of paper were used here, including some thin tissue paper with linework on the back.

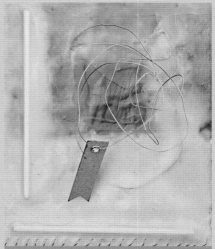

All sorts of objects can be embedded. Here, I used bamboo skewers and copper wire.

TECHNIQUE 6: PAPER

Paper of all sorts can be used to create many different effects. Thin paper such as tissue paper and napkins have the advantage of making the background virtually disappear. Only the pattern remains visible. This technique can be used to add art mediums that aren't compatible with wax, such as chalk or ink on tissue paper. The patterned paper is then applied to the artwork. Napkins, which are usually 2 or 3 ply, can be reduced down to the single printed layer and added to artwork. Some napkins have elements that can be added in their entirety. Others have useful patterns that can be layered to great effect.

To add tissue paper to an art piece, arrange it directly on a warm encaustic surface and burnish it in place with the back of a spoon. Then, apply a layer of encaustic medium over the paper and fuse with a heat gun or torch.

Lightweight paper can also be used for embellishment. The white spaces may not disappear completely, but the paper does get more transparent once it has absorbed the encaustic medium. These papers behave well and are fun to layer.

To use lightweight paper in an art piece, it is probably best to first dip the paper in encaustic medium and then place the paper on a warmed encaustic surface. Burnish slightly or heat with the heat gun to stick the paper to the art piece. Apply a layer of encaustic medium to seal it in place and fuse.

Working with medium-weight and heavyweight papers is also possible, but it can be more of a headache. Heavier papers seem to have a mind all their own. Some will stick perfectly, while others refuse to lie flat. Great effects can be achieved using them, but this is definitely an advanced technique that requires patience. The technique is basically the same as with lightweight paper, except that heavier paper will require more coaxing to keep it flat and completely embedded.

TECHNIQUE 7: FOUND OBJECTS

All sorts of found objects can be incorporated into an encaustic collage. Leaves and sand work well and make interesting compositions. Larger items such as sticks and branches can be used, but it can be challenging to get them to adhere to the surface of the work. Large heavy items are best attached to the art board with wire or nails to ensure they stay put. Advanced planning is key with these types of items.

To incorporate small found objects, first apply a base of warm wax and then press the object into the warm wax. Add additional wax around or on top of the object and fuse into place. Keep adding wax and fusing until the object is secured to the artwork. For really small items such as sand, it may not be necessary to add an additional layer of wax.

ENCAUSTIC ARTWORK

With this encaustic project, my goal is to illustrate that a variety of techniques can be employed to accomplish a personal vision for an art piece. This piece is the result of a whole bunch of U-turns and dead ends that I managed to turn around into something that works for me.

1. When I first started this piece, I was paralyzed by a clean white board. To get beyond that, I first added some tissue paper from some old sewing patterns and some other paper that I had been saving for encaustic work. Now, I had a piece that was no longer pristine, but not something that was working. So, I placed a wide strip of painter's tape across the bottom half of the board. Above the line, I stippled some colored wax and built up the texture a bit using the dry-brush technique. I liked where that was going, so I selected another color and repeated the process below the tape. But this time, I created a finer texture in the wax.

2. I was happy with the new direction of my piece, so I added some interest to the texture by varying the color in the wax layers. The different colors helped to give the texture dimension. I gave the texture a final fusing and removed the tape. Now, I had a wide swath of white to contend with.

3. The wax on either side of the taped-off area was thicker than the area I was about to work on, so I decided to use some tissue paper to add pattern and color. Doing this was easier than trying to work precisely in a relatively narrow depressed area. I drew circles using oil pastels onto some thin white tissue paper. Once I had the colors and shapes the way I wanted, I cut them into a strip that was narrower than the white area on my art piece. I warmed the wax and placed my colored circles onto the white area with the colored side facing the wax. I used my finger to burnish the paper into place and then added a couple layers of clear encaustic medium. The paper virtually disappeared, and the circles showed through. There was no risk of smudging because the pastels were sandwiched between the paper and the wax. This technique works well for incorporating art materials that could not be used directly on the wax.

4. Next, I decided to add a square of corrugated cardboard. Ideally, I should have masked off an area for the square before I started texturing, but at the time, I didn't know that. So, I used my clay tools to carve out an area for the cardboard. I scraped away the texture until it was the same level as the circle layer.

5. To attach the cardboard, I quickly applied a layer of encaustic medium to the board and the back of the cardboard and held the cardboard in place until it had cooled sufficiently. I then added encaustic medium to the sides to secure it.

6. The corrugated cardboard looks interesting when waxed. The color deepens to a pleasant shade of brown and the corrugations become slightly visible. I added a thin piece of paper with some faint writing on it to the surface of the cardboard and topped it with another couple layers of wax. Finally, I gave the whole piece a final fusing, polished up the smooth areas with a soft cotton cloth, and I was done.

BATIK

Batik is an ancient art form that is at least 2,000 years old. It is essentially a resist technique, which means that color on cloth is protected from additional dye by applying a wax covering.

Each color builds on the sum of the previous colors. It is an art form that can take years to perfect, requiring a vision for the end product and the patience to see the piece through all the necessary steps.

I like that batik can be as intricate or simplistic as time and skill will allow. Traditional batik often uses specialized tools, which can be difficult to find locally and require skill to operate. So, for my batik project, I decided to borrow from the Japanese Shibori dyeing technique. Shibori is the Japanese manual resist dyeing technique, which produces patterns on fabric. The fabric is manually folded, twisted, stitched, or compressed, making some areas impenetrable to dye to create interesting patterns.

THE WAX BLEND

The wax used in this technique is generally a blend of beeswax and paraffin. Depending on the ratios of the two, different effects can be achieved. The more paraffin that is used, the more the wax will crackle and allow some dye to seep through. Personally, I love this effect, so my wax blend will do a bit of that. Adjust the proportion of the wax blend as desired: more paraffin, more crackle; more beeswax, less crackle.

Also, learn from my mistake. Purchase new paraffin wax, rather than using the wax from recycled used candles. Although they may seem to be all paraffin, candles may contain oils and other fillers. The oils will not wash out easily and may leave a permanent grease stain on the fabric. In addition, the oils may make the wax more pliable, which is counter to why the paraffin is used in this technique in the first place.

THE DYE

For batik, I like to use specially designed cold-water dyes that do not require heat in the dyeing process. The grocery store dyes should work as long as they are prepared according to the directions and then allowed to cool. If the dye bath is too warm, it will melt the wax. Wetting the fabric with cold water first will allow the fabric to absorb the dye more evenly. Leave it in the dye bath until the desired tint has been achieved.

TEST THE DYEING TIME

I recommend using fabric scraps to test the dyeing time. Have several that can be pulled out at regular intervals, given a quick rinse, and tossed in the dryer to see how that color looks when dry. Consider setting a timer for more precise results. Once the dry sample is the desired color, remove the batik from the dye bath and rinse with cold water until the water is relatively clear. This is especially important with batik because any "free" dye remaining in the fabric may tint the areas where there is wax.

BASIC BATIK WAX RECIPE

MATERIALS

Beeswax (yellow or white)	9.6 oz	60%
Paraffin wax	6.4 oz	40%
Double boiler or electric skillet		
Wooden stir stick		
Nonstick mini bread pans		

Yield: 1 pound (455 g) batik wax

Athough any proportion of beeswax to paraffin can be used, my preferred wax blend follows. I usually make a good-size batch, mold it in small bread tins, and then simply melt the amount needed for a particular project. This recipe yields one pound (455 g) of batik wax, which is plenty for the following project, but actual coverage is dependent on how much of the fabric will be waxed.

1. Melt the waxes in a double boiler or electric skillet.

2. Stir to ensure it is well mixed.

3. Pour into mini bread pan molds until ready to use.

SHIBORI-INSPIRED BATIK FABRIC

I have chosen to use a relatively simple folding technique to create a pattern on the fabric. I used a basic white cotton bandana, but any lightweight cotton fabric will work. I will be using a 60 percent beeswax, 40 percent paraffin wax blend for this project. How much is needed will depend on the size of the electric skillet and how deep the wax needs to be. I kept the wax level at about ½ inch (1 cm).

MATERIALS

1 cotton bandana

1 tablespoon (15 ml) basic laundry detergent

1 tablespoon (15 g) washing soda (Sodium Carbonate – found in the laundry aisle)

1 batch basic batik wax (page 97)

Small electric skillet

Bamboo skewers

Rubber bands

Cold-water dye

Tub for dye vat

Stock pot or iron

Newspapers

Paper towels

1. First, make sure the bandana is ready to absorb the beeswax and dye by soaking it for an hour in a solution of water with some detergent and washing soda added. The idea is to remove all traces of residues, such as sizing, that might be on the fabric. Do not use any kind of softener in the wash or dry cycle. Rinse well and dry.

2. Next, prepare the wax. Add enough of the batik wax to the electric skillet to ensure it is ½ inch (1 cm) deep when melted. Slowly warm the wax until it is liquid. Set the skillet to a consistent temperature to make sure that the wax remains liquid, but not too hot, between 155°F and 175°F (68°C and 79°C).

3. Set up the dye bath following the dye manufacturer's instructions. I like to use my large stainless steel pots, which can be easily cleaned afterward.

4. Now, fold the fabric. I used a series of folds that form a hex pattern. The idea is to put folds into the fabric and then dip the folded fabric into the wax, creating the resist. Any pattern will work.

To re-create my pattern, follow the diagram (opposite page) for the folding technique. It looks more complicated than it is. The more precisely the folds line up, the clearer and more defined the wax resist lines will be.

5. Once the bandana has been folded, use the bamboo skewers, one on either side of the folded fabric triangle, and join them with rubber bands on either side. The skewers hold the folded fabric and keep the folded edges aligned. Now, the fabric is ready for the wax.

6. To create the pattern I made, first dip the long side into the wax, holding the fabric in the wax for a couple seconds. Check to make sure wax has reached all the areas uniformly. Turn the triangle around and dip the point on the opposite side.

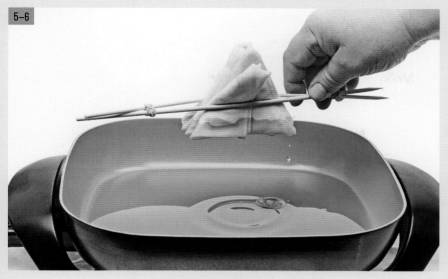

5–6

7

8

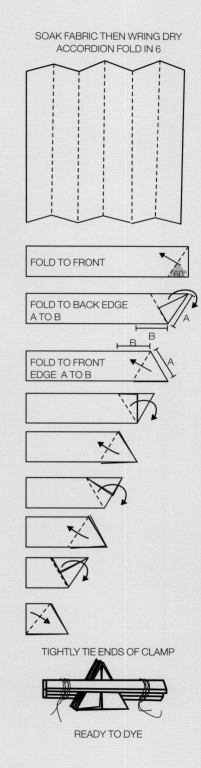

SOAK FABRIC THEN WRING DRY
ACCORDION FOLD IN 6

FOLD TO FRONT 60°

FOLD TO BACK EDGE
A TO B

FOLD TO FRONT
EDGE A TO B

TIGHTLY TIE ENDS OF CLAMP

READY TO DYE

7. Once the wax has cooled slightly, undo the bamboo skewers and unfold the fabric. Wet the fabric in cold water, either under the tap or in the sink, scrunching the bandana to crinkle the wax. Wetting the fabric first will allow the dye to color the fabric more uniformly. Transfer the fabric to the dye bath. Leave in the dye bath until the desired color is achieved (see "Test the Dyeing Time" on page 96). Rinse the fabric until the water runs clear. This will take several rinses. Once there is no more dye coming out of the fabric, fix the dye according to the manufacturer's directions and allow the fabric to dry.

8. Remove the wax using one of two methods. The first option is to boil the fabric in hot water. The wax will melt and float to the top. Once the water is cool, the wax can be skimmed from the surface of the water and the fabric lifted out and dried.

The second method is to use a hot iron and melt the wax into paper towels. First, cover the work surface with multiple layers of newspaper to protect it and then pile on several layers of paper towels. Finally, place the batik piece over the paper towels and cover the fabric with another couple layers of paper towels. As you iron, keep replacing the paper towels until all the wax is removed.

CHAPTER 6
ALCHEMY IN THE KITCHEN

From the time people discovered bees and their colonies, bees have always been used as a source for food. Initially, honey and honey comb were simply eaten as they were found. Over time, people learned where the bees stored the honey, pollen, and brood and harvested more selectively. As we developed pots for storage and cooking, honey became something that could be stored and used as needed. Foods were created that incorporated bee-related materials. Today, we have modern systems in place that enable us to separate honey, beeswax, and other bee-related products.

It is impossible to write a book about honey without including some food recipes. However, this book is not intended to be a cookbook. I have tried to showcase recipes that utilize honey, beeswax, and pollen in transformative ways. Also, because I dislike recipe books that use part of something, such as egg yolks in Cannelé Bordelais (page 106) for example, without suggesting ways to use the remainder, I have also included some recipes to make use of what is leftover. Those egg whites not used in Cannelé Bordelais can be used in the recipe for Chunky Buckwheat Granola (page 111).

COOKIES AND CANDIES

Cookies and candy are the perfect vehicles for showcasing honey. Honey cookies, energy bites, and peppermint patties come together quickly and are easy to make, even for novices. Caramels made with honey are a bit more advanced, but can be very rewarding once the technique is mastered.

HONEY COOKIES

These sweet and chewy cookies are simple, easy to make, and delicious.

INGREDIENTS

¹/₂ cup (225 g) butter, softened

¹/₂ cup (115 g) dark brown sugar, packed

¹/₂ cup (170 g) honey

1 egg

1¹/₂ cups (188 g) all-purpose flour

¹/₂ teaspoon baking soda

¹/₂ teaspoon salt

¹/₂ teaspoon cinnamon

Baking sheet

Yield: Makes approximately 15 cookies

1. Preheat the oven to 375°F (180°C, or gas mark 4).

2. Beat together the butter, brown sugar, honey, and egg in a medium bowl until smooth, scraping the sides occasionally. Stir in all the remaining ingredients.

3. Drop the dough by spoonful onto a greased or lined baking sheet. Bake for about 7 to 10 minutes or until the cookies are set and the edges are beginning to brown. The cookies will still look shiny when they're done.

4. Remove them from the baking sheet, place on a cooling rack, and allow to cool completely. These are best enjoyed fresh, but if necessary they will keep for several days in an airtight container.

ENERGY BITES

These bites are easy to make and provide the right amount of energy and nutrition in a single portion. This recipe can be adapted to include whatever ingredients are readily available. I like to include cacao nibs in my personal batches, but the harder texture has been misinterpreted as a nut shell, so I usually leave them out when making them for others. Also, I recommend storing in an airtight container in the refrigerator; they seem to hold up better when kept cool.

INGREDIENTS

2 cups (160 g) oats

1 cup (weight will vary) seeds (I like a mix of sunflower and pepitas.)

1/2 cup (weight will vary) nuts, chopped (I use pistachios.)

1/2 cup (weight will vary) dried fruit chopped if needed (I buy a mixed berry blend of blueberries, cherries, and cranberries.)

2 tablespoons (44 g) flax seed, ground

2/3 cup (230 g) honey

1/2 to 3/4 cup (130 to 195 g) nut butter (I prefer almond or cashew.)

1 tablespoon (15 ml) vanilla extract

4 tablespoons (36 g) pollen

Medium bowl

Small bowl

Wooden spoon

Yield: Makes approximately 30 energy bites

1. Measure all the dry ingredients into a medium bowl. Set aside.

2. Measure the honey and nut butter into a small bowl. Warm the mixture slightly to make it easier to stir. Add vanilla extract and pollen. Stir to combine.

3. Add the honey nut butter mixture to the dry ingredients and mix thoroughly.

4. Form into bite-size balls about 1 1/2 inches (4 cm) in diameter. Store in an airtight container in the refrigerator. They will keep for several weeks if stored in the refrigerator.

HONEY CARAMELS

I admit it. I have a weakness for caramel. Growing up, I loved Marathon candy bars, which were made with super hard, super chewy caramel—real filling pullers. Most of the caramels available these days are super soft and hardly chewy. I tried to find a halfway point with this recipe, a happy medium between the caramel of my youth and the soft caramels of today. I have also added a chocolate option, which is delicious.

INGREDIENTS

1 cup (235 ml) heavy cream

1 vanilla bean, split lengthwise

3 tablespoons (15 g) unsweetened cocoa powder (optional)

1 1/3 cups (267 g) sugar

2/3 cup (230 g) honey

1 stick (4 ounces, or 112 g) of unsalted butter, softened and cut into chunks

1 teaspoon coarse sea salt

Baking dish, 9 inch x 9 inch (23 cm x 23 cm)

Wax paper

Small saucepan

Large saucepan

Whisk

Candy thermometer

Sharp knife

Cutting board

Yield: About 35 caramels

1. Line the baking dish with wax paper, leaving long overhangs on two sides.

2. In a small saucepan, combine the cream and split vanilla bean and simmer over low heat for 10 minutes. Remove the vanilla bean, scrape out seeds, and add to the cream. Add the cocoa powder, if desired, and stir to combine. Keep warm on low heat.

3. In a large saucepan, combine the sugar and honey. Without stirring, dissolve the honey and sugar mixture over medium heat until smooth and melted. Continue heating the mixture until it has darkened to a deep caramel color, about 5 minutes. Watch carefully—sugar burns quickly!

4. Remove from the heat and whisk in the chunks of butter one at a time. Once all the butter is added, whisk in the hot vanilla cream mixture.

5. Bring the pot to a boil over medium heat and continue to boil until the mixture reaches the hard ball stage (see sidebar). Remove from the heat and pour the caramel into the prepared pan.

6. Place the pan in the refrigerator for about 10 minutes to set up slightly and then sprinkle top of caramels with sea salt. Let the caramels set up at room temperature for about an hour or until totally cooled.

7. To remove from the pan, gently pull on the wax paper and remove the caramel block from the pan. Cut into squares with a sharp knife and wrap in small pieces of wax paper.

8. Keep the wrapped caramels in an airtight container to prevent them from attracting moisture and getting gummy on the outside. Assuming they don't get eaten first, they should keep for several weeks.

PEPPERMINT PATTIES

Peppermint patties are quick and easy to make and require only a few ingredients. I used an artisan chocolate bar with 72 percent cacao from my local grocery, but any bittersweet chocolate will work. I have also seen these made with unsweetened chocolate, as the honey adds plenty of sweetness, but I find the flavors a bit too disparate for my taste. Also, this works best with solid honey: the stiffer, the better.

CANDY-MAKING STAGES

To test candy stage without a thermometer, a small amount of the cooked sugar syrup is dropped into a glass of very cold water and observed. The chart below outlines the various stages.

Thread: syrup spins a soft, loose, short thread
 230°F to 235°F (110°C to 113°C)

Soft Ball: syrup forms a soft, pliable, sticky ball
 235°F to 240°F (113°C to 116°C)

Firm Ball: syrup forms a firm, but still pliable, sticky ball
 245°F to 250°F (118°C to 121°C)

Hard Ball: syrup forms a hard, sticky ball
 250°F to 265°F (121°C to 129°C)

Soft Crack: syrup forms longer strands that are firm, remain pliable
 270°F to 290°F (132°C to 143°C)

Hard Crack: syrup forms stiff strands that are firm and brittle
 300°F to 310°F (149°C to 154°C)

Caramel: syrup forms hard strands that are firm and brittle
 320°F to 335°F (160°C to 168°C)

INGREDIENTS

3.5 to 4 ounces (100 to 115 g) bittersweet chocolate

3 tablespoons (60 g) solid honey

1/4 teaspoon peppermint oil (food-grade)

Double boiler

1/2 teaspoon measuring spoon

Silicone mini-muffin mold (makes 12)

Small bowl

Spoon

Candy foil

Yield: 12 patties

1. Melt the chocolate in a double boiler. Once melted, drizzle about 1/2 teaspoon of the chocolate into the bottom of each silicone mini-muffin cup. Use the spoon to spread the chocolate up the sides a bit and allow to harden.

2. In a small bowl, mix the honey and peppermint oil.

3. Once the first layer of chocolate has hardened, spoon a dollop of the honey mixture in the center of each cup and top with the rest of the melted chocolate. I usually start drizzling around the outside and work toward the middle. Chill thoroughly and pop out of the molds. I like to wrap mine in candy foil for a decorative touch. Store in an airtight container. Keeps for several months.

DESSERTS

Desserts are a traditional way to use products from the hive. I have gathered a collection of various recipes that utilize not just honey, but beeswax as well.

CANNELÉ BORDELAIS

My love affair with Cannelé began when I was researching uses for beeswax for my last book, *Beeswax Alchemy*. Whole treatises have been written chronicling the quest for the perfect Cannelé. There was no way I had time to perfect something so elusive for my last book! I was very disappointed.

But, once the book was written and I had some free time, I decided to delve into the Cannelé world myself. After all, a pastry that is dark and crunchy on the outside and super moist, almost liquid, on the inside seemed to be the perfect pastry for me. And, best of all, it includes beeswax, which is mixed with butter and used as a mold grease. As the Cannelé bakes, the beeswax adds a warm honey-like scent and flavor, while adding a nice crunch. I have seen recipes that also use honey in place of the sugar, but for simplicity, I stuck with sugar for this recipe.

These pastries come from the Bordeaux region of France, which is a region known for its wine. One theory is that this recipe came about when winemakers, who used egg whites to clarify wine, were looking for uses for leftover egg yolks. It seems plausible to me.

The Cannelé are traditionally baked in small copper molds. The individual molds are taller than they are wide and they are fluted. Although I would love to try making them with copper molds, the molds are very expensive. Cheaper alternatives are available. I purchased some aluminum single molds and a silicone mold with 6 cavities. Although I was happy with both, the aluminum molds yielded a Cannelé that was more crunchy and browned.

INGREDIENTS

Batter:

2 cups (475 ml) whole milk

1 1/2 ounces (42 g) unsalted butter

1 vanilla bean, split with seeds scraped

3/4 cup (150 g) sugar

3/4 cup (94 g) flour

1/4 teaspoon salt

2 large eggs

2 large egg yolks

1/4 cup (60 ml) dark rum

Mold Grease:

1 tablespoon (14 g) beeswax

1 tablespoon (14 g) unsalted butter

Small saucepan

Medium bowl

Small bowl

Wooden spoon

Container with airtight lid

Cannelé molds (either copper, aluminum, or silcone)

Small, heatproof container

Clean brush for mold grease

Baking sheet

Yield: Makes approximately 15 Cannelé

CANNELÉ TIPS:

The less air that is incorporated, the better. Stir, don't whip.

Allow 2 full days of rest in the refrigerator.

Don't crowd the Cannelé on the baking sheet.

Turn the baking sheet.

They are best enjoyed fresh out of the oven.

This recipe is the perfect starter recipe. I add orange zest to the milk when I make mine, but all sorts of flavors can be added to tweak the recipe. Try some lavender blossoms, star anise, or even coffee.

1. Heat the milk, butter, and vanilla bean and seeds in a saucepan over medium heat until the butter is melted and it is at a low simmer. Remove from heat and allow to cool for a bit. Remove the vanilla bean.

2. In a medium bowl, stir together the sugar, flour, and salt. Set aside.

3. In a small bowl, stir together the eggs and egg yolks, taking care not to incorporate too much air. Temper the eggs by adding small amounts of warm milk to the eggs and stirring before adding more milk. The idea is to raise the temperature of the eggs without cooking them. Once about half the milk is stirred into the eggs, add the remaining milk and egg mixture to the sugar and flour mixture. Stir just enough to incorporate. Add the rum and pour the mixture into an airtight container and refrigerate.

4. Leave the mixture to rest in the refrigerator for at least 2 full days, stirring occasionally. Allow to come to room temperature for an hour before baking.

5. When ready to bake, preheat the oven to 475°F (240°C, or gas mark 9) and prepare the molds. I use the same technique for both silicone and aluminum molds. First, melt the beeswax and butter in a small, heatproof container. To coat the molds, heat the molds slightly. Brush the beeswax/butter mixture in a thin layer inside the molds and pop into the freezer to cool.

6. Place the molds on a baking sheet, allowing for plenty of air space around each mold. Give the batter a gentle stir and pour into the waiting molds. Fill the molds about 3/4 full.

7. Once the oven is hot, carefully transfer the baking sheet to the oven and immediately lower the temperature to 425°F (220°C, or gas mark 7). Bake for 15 minutes. Lower the baking temperature to 375°F (190°, or gas mark 5) for another hour or so. Bake until the outside is medium to dark brown (but not burnt). Take the baking sheet out of the oven and allow the Cannelé to rest for 10 minutes before unmolding them onto a cooling rack.

HONEY CITRUS TEA CAKES

Tea cakes are simple cakes originally intended to be served with tea. They come together quickly and are relatively foolproof. Although this tea cake can be made without the citrus, it would be a shame because the flavors meld beautifully with the honey. I bake this in a loaf pan, but traditionally, it is baked in a round cake pan.

INGREDIENTS

2 cups (260 g) + 2 tablespoons (16 g) all-purpose flour

2¼ teaspoons baking powder

½ teaspoon salt

Freshly grated zest and juice of 2 blood oranges

Freshly grated zest and juice of ½ of a lemon

4 large eggs, at room temperature

½ cup (170 g) honey

¾ cup (175 ml) mild extra virgin olive oil

½ cup (120 ml) milk

Grater

Citrus juicer

8-inch (23 cm) loaf pan

Parchment paper

Small bowl

Medium bowl

Whisk

Wooden spoon

Yield: Makes one 8-inch (20 cm) loaf

1. Preheat the oven to 350°F (180°C, or gas mark 4). Line a loaf pan with a piece of parchment paper long enough to hang over the sides (this acts a handle to easily lift the baked loaf from the pan).

2. In a small bowl, whisk together the flour, baking powder, salt, blood orange zest, and lemon zest.

3. In a medium bowl, whisk together the eggs, honey, olive oil, and blood orange and lemon juices. Whisk vigorously until smooth and there are no lumps. Combine the milk and the flour mixture and stir until just blended and there are no visible lumps of flour.

4. Scrape the batter into the prepared loaf pan. Bake for 50 minutes or until a deep golden color and the cake springs back when gently tapped with your finger.

5. Let the cake cool completely before slicing. Wrap any leftover cake tightly in parchment paper and enjoy within 2 days.

MANGO SHRIKHAND

This dessert has its origins in Northern India. Shrikhand is a sweetened, strained yogurt often served with added fruits, nuts, and spices. I usually make this when I need whey for lacto-fermentation (see recipe, page 127). It is perfect for hot summer days!

INGREDIENTS

3/4 cup (180 g) strained yogurt (approximately 2 cups [460 g] unstrained)

1 to 2 tablespoons (15 to 28 ml) milk

Saffron, a few threads, crushed

1/4 cup (85 g) honey (if mangoes are super sweet, start with less)

1/4 teaspoon green cardamom powder

1/4 to 1/2 cup (62 to 125 g) mango puree

6 to 8 pistachios (or other nuts such as, almonds or cashews) chopped finely, optional

Medium bowl

Small bowl (microwave safe)

Wooden spoon

Yield: Makes 2 to 3 servings (1/2 cup, or 120 ml, each)

1. Pour the strained yogurt into a medium bowl and set aside.

2. Pour the milk into a small, microwave-safe bowl and warm to a temperature of approximately 120°F (49°C). Add the saffron and mix. While still warm, add the honey and stir to combine. The warmth of the milk should help to soften the honey, allowing it to mix with the cool yogurt.

3. Add the milk and honey mixture, cardamom powder, and mango puree to the strained yogurt. Stir gently until it is completely mixed.

4. Spoon the mixture into dessert dishes and chill. If desired, top with the chopped nuts right before serving. Best enjoyed within a day or two.

CHUNKY BUCKWHEAT GRANOLA

Granola is my go-to snack in my workspace. It is easy to eat and substantial enough to tie me over until dinner. It is also a good way to use up leftover egg whites. Although not required, for those like me, who prefer a chunky granola, using egg whites is key to creating nice big chunks.

INGREDIENTS

3 cups (240 g) rolled oats (gluten-free if necessary)

1 cup (240 g) buckwheat

1 1/2 cups (90 g) coconut flakes

1/4 cup (52 g) chia seeds

1/4 cup (36 g) coconut sugar

1 cup (135 g) hazelnuts (Walnuts are also delicious.)

1/3 cup (75 g) coconut oil

1/3 cup (115 g) honey

1 teaspoon vanilla extract

1/2 teaspoon fine grain sea salt

1/2 cup (40 g) cocoa powder (organic, fair-trade if possible)

2 to 3 egg whites (optional)

Large bowl

Knife

Cutting board

Small saucepan

Wooden spoon

Small bowl

Whisk

Spatula

Baking sheet

Parchment paper

Yield: 8 cups (900 g)

1. Preheat the oven to 350°F (180°C, or gas mark 4).

2. In a large bowl, combine the oats, buckwheat, coconut flakes, chia seeds, and coconut sugar. Roughly chop the nuts and add them to the mix.

3. In a small saucepan over low-medium heat, melt the coconut oil. Add the honey, vanilla, salt, and cocoa powder. Whisk to combine until smooth.

4. Whisk the egg whites in a small bowl until fluffy.

5. Pour the honey/oil mixture over the dry ingredients and fold with a spoon to coat completely and evenly. Add the whipped egg whites and mix thoroughly.

6. Spread the mixture out in an even layer on a lined baking sheet and press firmly with the back of a spatula to ensure that the mixture is compact. Bake for 15 to 20 minutes.

7. Remove from the oven, flip the granola in large chunks, and place back in oven to bake for another 10 minutes, stirring every 3 to 4 minutes until toasted and fragrant. The dark color of the granola with cocoa powder makes it hard to tell if it is cooked or not, so go by smell. Another good way to test it is by tasting a hazelnut, which takes the longest to cook—it should taste nutty and pleasantly roasted. Store granola in an airtight container for up to several months.

ICE CREAM

Ice cream and honey are a match made in heaven. Homemade ice cream often gets too hard once it freezes completely. By using honey instead of sugar, the freezing temperature of the mixture is lowered, changing the crystal structure to yield an ice cream that stays softer and scoopable. It is best made in a manual or electric ice cream maker.

HONEY ICE CREAM

Technically, this recipe is a frozen custard and is a nod to my home city of Milwaukee, which is known for frozen custard. In my head, it is ice cream and its all good!

1. Place the container in which you plan to store the finished ice cream into the freezer to chill. In a medium saucepan, combine the cream, milk, and honey. Warm over medium heat until barely simmering, stirring frequently. Remove from heat and cover. Set aside.

2. In a medium bowl, whisk the egg yolks. Temper the egg yolks by slowly pouring some of the hot cream into the yolks while whisking to raise the temperature and keep the egg yolks from cooking. Then, pour everything back into the saucepan.

3. Heat the mixture over medium heat, stirring constantly and scraping the bottom as you stir. While the custard heats, stir in the salt and vanilla extract. Gently cook until the mixture thickens enough to coat the back of a wooden spoon, about 4 minutes.

4. Pour the custard through a fine-mesh strainer into a clean bowl. Place the bowl in an ice bath and stir the custard occasionally until it is cool, about 20 minutes. Cover and refrigerate for at least 3 hours or overnight.

5. Pour the chilled custard into the ice cream maker and follow the manufacturer's instructions.

6. After the ice cream has reached the desired consistency, scrape it into the prechilled container, cover, and place in freezer. I prefer to eat this right away, but it can be stored in an airtight container in the freezer for several weeks.

INGREDIENTS

1 1/2 cups (355 ml) heavy cream

1 1/2 cups (355 ml) whole milk

1/3 cup (115 g) buckwheat honey or slightly more of a mild-flavor honey

5 large egg yolks

Pinch of salt

1/2 teaspoon vanilla extract

Medium saucepan

Wooden spoon

Medium bowl

Whisk

Fine-mesh strainer

Clean bowl

Cling wrap

Ice cream maker

Tight-sealing container for finished ice cream

Yield: Approximately 1 pint (475 ml)

BEESWAX ICE CREAM

Why put beeswax in ice cream? It effectively raises the melting point of the ice cream, making it less messy to eat on really hot days. The recipe is similar to the honey ice cream, but the addition of beeswax means more egg yolks are needed to help emulsify the wax into the cream.

INGREDIENTS

2 cups (475 ml) heavy cream

1 cup (235 ml) whole milk

1/3 cup (115 g) buckwheat honey or slightly more of a mild-flavor honey

7 large egg yolks

Pinch of salt

1/2 teaspoon vanilla extract

1/2 cup (115 g) beeswax, melted

Medium saucepan

Wooden spoon

Medium bowl

Whisk

Blender

Fine-mesh strainer

Clean bowl

Cling wrap

Ice cream maker

Tight-sealing container for finished ice cream

Yield: Approximately 1 pint (475 ml)

1. Place the container in which you plan to store the finished ice cream into the freezer to chill. In a medium saucepan, combine the cream, milk, and honey. Warm over medium heat until barely simmering, stirring frequently. Remove from the heat and cover. Set aside.

2. In a medium bowl, whisk the egg yolks. Temper the egg yolks by slowly pouring some of the hot cream into the yolks while whisking to raise the temperature and prevent the egg yolks from cooking. Then, pour everything back into the saucepan.

3. Heat the mixture over medium heat, stirring constantly and scraping the bottom as you stir. While the custard heats, stir in the salt and vanilla extract. Gently cook until the mixture thickens enough to coat the back of a wooden spoon, about 4 minutes.

4. Remove from the heat and slowly whisk the melted beeswax into the hot custard. Pour the entire contents into a blender and blend on high for 30 seconds. Strain the mixture into a clean bowl through a fine-mesh strainer to capture any wax solids that have not been incorporated. Place the bowl in an ice bath and stir the custard occasionally until it is cool, about 20 minutes. Cover and refrigerate for at least 3 hours or overnight.

5. Pour the chilled custard into the ice cream maker and follow the manufacturer's instructions.

6. Once the ice cream has reached the desired consistency, scrape the finished ice cream into the prechilled container, cover, and place in freezer. I prefer to eat this right away, but it can be stored in an airtight container in the freezer for several weeks.

VARIATIONS

These basic recipes can be tweaked in many ways. Try infusing the hot cream with herbs or spices or adding fruits and nuts to the ice cream maker.

IDEAS FOR INFUSING
Chamomile blossoms, lavender blossoms, rose petals, rosemary, thyme, basil, mint, cardamom, cinnamon, star anise, lemon verbena, anise hyssop, rose geranium

FLAVOR COMBINATIONS
Basil and peach, chamomile and lemon zest, rose and strawberry, basil and blackberry, mint and lemon, lavender and blueberry

APPETIZERS

This category should probably be called "Everyday Recipes," because most of the recipes make their way into regular rotation in my house, especially the bread. I always have a loaf in the works. Even the Pistachio and Honey Chevre log, which is a staple at gatherings, is something I also make for the family on occasion.

PISTACHIO AND HONEY CHEVRE LOG

This is one of my favorite last-minute go-to recipes if I have forgotten to make something for a get-together at a friend's house. The sweetness of the honey is a nice counterpoint to the tanginess of the chevre and the pistachios add a pleasant crunch. I use a fig jam, but almost any jam would be nice, even something with a little bit of heat.

INGREDIENTS

1 log (10 ounces, or 280 g) of chevre goat cheese

1/4 cup (85 g) honey

2 tablespoons (40 g) fig jam

1/8 to 1/4 cup (15 to 31 g) shelled, chopped pistachios

Serving plate

Small microwave-safe bowl

Spoon

Yield: Serves 10 to 12, or more as an appetizer

1. Place the chevre cheese log on serving dish.

2. Warm the honey and jam in a small bowl in the microwave until the preserves are melted and the honey and jam can be easily combined.

3. Drizzle the honey-jam mixture over the goat cheese log and sprinkle with chopped pistachios.

4. Serve with crackers or crusty bread.

BREAD WITH HONEY BUTTER

What could be better than fresh baked bread with butter, especially honey butter? In my world, nothing. Although this bread uses only a bit of honey, it makes the perfect vehicle for honey butter.

Ever since I read Michael Pollan's *Cooked*, I have been obsessed with sourdough bread. I am not gluten intolerant, but I do believe that fermenting the dough before baking makes it easier for the body to digest the wheat properly. To that end, I have tried multiple times—unsuccessfully—to create my own sourdough starter. Having failed at that, I recently dusted off a technique I used many years ago, which was counter to everything I knew about breadmaking at the time: mix yeast with water, add flour, knead, set in warm place for a couple hours, shape and put in loaf pan, final rise, and then bake. But that technique produced bread-shaped rocks. The epiphany for me was the concept of the slow-rise loaf. This technique used a fraction of the yeast and allowed the bread to rise slowly in a cool location overnight before baking the following day, having slightly soured in the process. Success!

Although this method is not new or quick, I have made some of the best breads since I began to using it. I start with a pre-ferment, sometimes known as a *biga* or *poolish*, and the following day, I add to it to make the actual dough. It often takes 3 days from start to finish, but requires very little actual hands-on work. I bake the bread in a cast iron Dutch oven. In this self-contained, moist environment, the bread develops a nice, crunchy crust.

RUSTIC DUTCH OVEN BREAD

1. To make the pre-ferment, stir all of the pre-ferment ingredients together to make a thick, wet mixture. Cover with plastic wrap and let it rest for at least 2 hours. For best flavor, let the starter rest longer or overnight.

2. To make the dough, stir the pre-ferment with a spoon and then add the water, yeast, honey, 3 1/2 cups (480 g) of the flour, and the salt. Mix or knead the dough, just until the ingredients are all incorporated. The dough should be a slightly shaggy, messy dough. Cover with a towel or plastic wrap and let it rest for 30 minutes to allow the flour to absorb the water and then knead it again. It should now be more cohesive and a bit smoother. Knead the dough, adding more flour if needed, to make a soft dough.

3. Place the dough in a lightly greased bowl, cover with lightly greased plastic wrap, and allow it to rise until almost doubled in a cool spot or in the refrigerator.

4. Carefully work the dough into one large loaf, trying not to deflate the dough completely. Dust a piece of parchment paper with cornmeal or flour. Gently place the dough on the parchment paper, seam side down, and cover with greased plastic wrap. Allow to rise in a warm spot until it rises by 50 percent or more.

5. Put the Dutch oven inside the oven and preheat both to 425°F (220°C, or gas mark 7). The pot may take a bit longer to heat up than the oven itself.

6. When the dough is ready, remove the pot from the oven. Pick up the parchment paper and dough together and place it directly into the pot. Slash or cross-hatch the bread with a sharp knife. Cover the pot with the lid and place in oven.

7. Immediately reduce heat to 375°F (190°C, or gas mark 5) and bake for 30 minutes. Remove the lid and bake an additional 20 to 30 minutes or until the bread is baked through. The internal temperature should be at least 190°F (88°C). Remove the bread from the Dutch oven and place on a wire rack to cool. Resist the urge to cut into the bread while it is still hot. The loaf is best enjoyed fresh but cool. It will keep for a couple days in a plastic bag.

INGREDIENTS

Pre-Ferment:

1 cup (235 ml) cool to lukewarm water (90°F to 100°F [32°C to 38°C])

1/2 teaspoon active dry yeast

1 1/4 cups (171 g) bread flour

1/4 cup (31 g) all-purpose flour or whole wheat flour

Large bowl

Wooden spoon

Plastic wrap

Dough:

Pre-ferment from above

1 cup (235 ml) water (100°F to 115°F [38°C to 46°C])

3/4 teaspoon active dry yeast

2 tablespoons (40 g) honey

3 1/2 to 4 cups (480 to 548 g) bread flour

2 teaspoons salt, or to taste

Plastic wrap

Cornmeal or flour

Parchment paper

Dutch oven

Sharp knife

Yield: 1 loaf (approximately 2 pounds, or 900 g)

HONEY BUTTER

INGREDIENTS

1 pound (455 g) butter

$1/4$ cup (85 g) honey

Knife

Medium bowl

Mixer

Parchment paper or plastic wrap

Yield: 1 pound (455 g) of honey butter

1. Cut the butter into chunks and add to the bowl. Mix the butter with mixer at low speed until it has loosened up and is easily workable.

2. Add the honey and mix at medium speed until well combined.

3. Spoon onto parchment paper or plastic wrap to form a log and refrigerate for several hours or until needed.

Make the honey butter extra special by adding $1/2$ teaspoon of ground cinnamon and $1/2$ teaspoon of vanilla extract along with the honey.

BASIL HONEY RICOTTA TARTINE

I fell in love with tartines, or French open-faced sandwiches, when I toured France. It seemed that every corner bistro had at least a couple on the menu. Served on a crunchy slice of sourdough bread, the toppings were usually simple, but elegant. This is one recipe I come back to time and time again. The recipe as shown utilizes the whole loaf of bread, but the recipe can easily be scaled down to a slice or two for a light lunch.

INGREDIENTS

1 loaf of crusty sourdough bread, cut into $3/4$- to 1-inch (2 to 2.5 cm) slices

1 cup (250 g) whole milk ricotta

2 lemons, zested

1 cup (24 g) sweet basil, larger leaves coarsely chopped

1 large clove of garlic, peeled

$1/2$ to 1 cup (170 to 340 g) mild honey

Microplane or zester for lemons

Grill pan or grill to toast bread

Yield: About 10 slices, depending on size and shape of loaf

1. Toast the bread slices on a grill or on the stove top in a grill pan for about 2 minutes per side. The bread surfaces should be toasted to a light to medium brown.

2. Rub the garlic clove across one side of the toasted bread.

3. Spread a layer of ricotta on the bread, add the basil, and dust the bread slices with lemon zest.

4. Right before serving, drizzle the honey over the top. Consume immediately.

ACCOMPANIMENTS

This is probably my favorite food section. It showcases the versatility of honey and its uses. Most of them come together easily and elevate the foods to which they are added.

HONEY MUSTARD

I love mustard! Mustard is easy to make, but it can be a bit tricky to strike the perfect balance of flavors. Although I love a spicy mustard, for a honey mustard, I prefer using mostly yellow mustard seeds (about 90 percent yellow, 10 percent brown). This mustard is made with no preservatives, so I like to make it in small batches in a wide-mouth pint-size (475 ml) canning jar. That way, it can be made and stored in the same container. This recipe calls for whole seeds instead of powder, which results in a course, slightly grainy mustard.

INGREDIENTS

¼ cup (44 g) yellow mustard seeds (or a mixture of yellow and brown mustard seeds)

¼ cup (60 ml) water

2 tablespoons (28 ml) apple cider vinegar

¼ teaspoon salt

2 to 4 tablespoons (40 to 85 g) honey (I prefer a mild honey.)

Wide-mouth pint-size (475 ml) canning jar

Immersion blender

Measuring cups and spoons

Yield: 1 pint (475 ml)

1. Measure the mustard seeds into the pint-size (475 ml) canning jar. Add the water and let sit for a couple minutes. Add the vinegar, cover the jar with a lid, and refrigerate overnight.

2. By the following day, the seeds will have soaked up most of the liquid. Use an immersion blender to puree the contents of the jar as much as desired. Add the salt and honey and mix well.

3. Add the lid and refrigerate the mustard for several days, allowing it to mellow a bit before assessing the flavor. Will keep for several months in the refrigerator.

ACHIEVING THE RIGHT FLAVOR

Mustard seeds come in either yellow (mild) or brown (spicy). The seeds in and of themselves don't really have any heat; it is only when the seeds are combined with a liquid that a chemical reaction releases the enzymes that make the mustard pungent. The acidity of the vinegar slows this chemical reaction. Using only water results in a mustard that is extremely pungent at first, but one that loses its pungency more quickly than a mustard with vinegar.

Also, mustard seeds are naturally somewhat bitter. Grinding the mustard seeds releases that unpleasant bitterness. Much of the bitterness will mellow over the first couple days, so don't judge the mustard until it has had a chance to age. Time works wonders.

HONEY AVOCADO DRESSING

This dressing tastes great on anything, from salads to wraps and sandwiches. It is the addition of the avocado that makes it decadently creamy. I like to use the fermented honey garlic on page 130 in this recipe.

INGREDIENTS

½ cup (120 ml) grapeseed oil

2 tablespoons (40 g) honey or fermented honey garlic (page 130)

2 cloves of garlic

1 medium avocado, peeled, pitted, and chopped

¼ cup (60 ml) lime juice

¼ cup (4 g) chopped cilantro

Salt and black pepper to taste

Blender

Spatula

Airtight container

Yield: Makes approximately 1 cup (235 ml) dressing

1. In a blender, combine the oil, honey, garlic, avocado, lime juice, and cilantro and season with salt and pepper. Puree until smooth.

2. Use a spatula to transfer the dressing to an airtight container.

3. Refrigerate for up to 3 days.

HONEY VINAIGRETTE WITH POLLEN

This dressing works well on "stronger" greens such as romaine lettuce, fresh spinach, or other sturdy greens.

INGREDIENTS

1/4 cup (60 ml) extra-virgin olive oil

1/4 cup (60 ml) lemon juice

1/4 cup (60 ml) apple cider vinegar

2 tablespoons (30 g) honey mustard

1 1/2 tablespoons (14 g) bee pollen

1 clove of garlic, minced

1 to 2 teaspoons honey (depending on sweetness of honey mustard)

1/2 teaspoon cumin

1/2 teaspoon sweet paprika

Salt and pepper to taste

Pint (475 ml) jar or carafe with lid

Yield: 1 cup (235 ml)

1. In jar or carafe, mix all the ingredients together.

2. Refrigerate for several hours for the flavors to meld and for the pollen granules to break apart.

3. Mix well before serving.

4. Keeps for about 1 week in the refrigerator.

HONEY BARBECUE SAUCE

This barbeque sauce is a combination of sweet and tangy and is simply the best way to top your summer BBQ!

INGREDIENTS

1 cup (240 g) ketchup

1 cup (235 ml) white vinegar

2 tablespoons (40 g) molasses

1 cup (340 g) honey

1 teaspoon salt

1/2 teaspoon pepper

2 teaspoons dry mustard

1 teaspoon paprika

1 1/2 teaspoons garlic powder

1 1/2 teaspoons onion powder

Medium saucepan

Whisk

Airtight container

Yield: Serves 8 to 10

1. In a medium saucepan, whisk all ingredients and warm over medium heat. Simmer the barbecue sauce for 10 to 15 minutes.

2. Remove from the heat and allow to cool.

3. Transfer to an airtight container and store in the refrigerator until ready to use. Use within 1 month.

SMOKED HONEY

Sweet, sticky, smoky honey! The earthy aftertaste of smoked honey enhances both sweet and savory dishes. Cold-smoking retains the natural honey components and makes a small batch. Smoked honey is a unique gourmet ingredient that is ideal for gifts and sharing with your family and friends. Although this is best made with a smoker where it can be done cold, it can also be done using a grill.

INGREDIENTS

Honey

Smoking wood chips

Smoker or grill

Foil trays

Wooden spoon

Foil tray lids, foil, or plastic wrap

Airtight containers

1. Pour the honey into foil trays (ensure that the honey is no thicker than 1/2 inch [1 cm] for maximum exposure).

2. Place the foil trays onto the wire rack in the smoker or grill.

3. Cold smoke the honey for 30 minutes for smaller smokers or 60 minutes for larger smokers. Stir every 15 to 20 minutes.

4. Remove the trays from the smoker or grill.

5. Cover the trays with a lid, foil, or plastic food wrap and set aside (indoors) at room temperature for 24 hours.

6. Taste the smoked honey, mixing with nonsmoked honey if the smoked flavor is too strong for your liking.

7. Pour the smoked honey into airtight containers such as glass jars with lids.

8. This can be used immediately or stored at room temperature as with regular honey. Stir the honey before use.

CONVENTIONAL GRILL SMOKING

To use a grill to smoke honey, start with a small amount of hot coals placed on one side of the grill. Place soaked wood chips directly on the hot coals. The goal is to keep the grill at about 175°F (79°C). Place the pans on the grate on the opposite side from the coals and smoke the honey for 30 to 50 minutes.

IDEAS FOR USING SMOKED HONEY

Smoked honey is gluten free and suitable for adding flavor to vegetarian and meat dishes as a glaze, marinade, or sauce. Drizzle or brush on the honey during cooking or enjoy straight from the jar! Here are a few other ways to enjoy smoked honey:

Cheese board

Chicken glaze

Cocktails

Fruit salad

Grilled figs

Ice cream topping

Jerky marinade

Lamb tagine

Pancakes

Pineapple

Pork chops

Roasted carrots

Salad dressing

Smoked honey ice cream

Vanilla milkshake

Yogurt

FERMENTED FOODS

FERMENTING BASICS

Although fermented foods have been around for a long time, they have recently found renewed favor. Before the invention of refrigeration, fermenting food was one way of preserving foods. Today's renewed interest comes from a new understanding of how the fermentation process actually enhances the nutritional value of food and helps our bodies digest foods that otherwise might be difficult.

Sauerkraut, pickles, and other vegetables that are grown in, or directly on, the ground are covered in the bacteria that act as a starter to fermentation. Fruits and vegetables that grow more removed from the soil may not come with their own handy, built-in bacterial starter, so one is needed to kick off the fermentation process. That's where whey comes in to save the day.

Although whey is not the only starter that can be used, it is one I always have on hand. The reason to use whey is that it contains *Lactobacillus*, the same bacteria that turns milk into yogurt. *Lactobacillus* is also the bacteria found naturally on vegetables. The term *lacto-fermentation* defines things fermented using the *Lactobacillus* bacteria.

STRAINING YOGURT FOR WHEY

This is my method for straining yogurt. The yield is somewhat variable, depending on how long the yogurt is strained. In general, I figure 1 quart (950 ml) of regular yogurt will yield approximately 1 cup (235 ml) of whey. Separated whey can be refrigerated for several months until used.

When straining yogurt for whey, use a high-quality yogurt. Store brands are fine to use in a pinch, but read the label to ensure that the yogurt contains live, active cultures. Store brands also often add ingredients to homogenize and thicken the yogurt. If possible, avoid yogurts with added ingredients or better yet, use homemade yogurt. Although it is possible to strain some whey off Greek-style yogurts, they will yield very little because most of the whey was already strained off prior to packaging.

INGREDIENTS

Plain yogurt with active cultures (one quart [950 ml] of yogurt will yield about 1 cup [235 ml] of whey)

Large canning jar with lid

Clean, muslin fabric

Rubber band

Yield: Approximately 1 cup (235 ml) of whey

1. Stretch the muslin fabric across the top of the canning jar, allowing the excess to drape over the sides. Push the fabric down into the canning jar, making sure the bottom of the muslin is suspended high enough above the bottom of the jar to allow room for the whey that drains off the yogurt to accumulate. Hold the fabric in place by securing a rubber band around the top.

2. Add the yogurt to the jar and cover the jar with the lid. Place the jar in the refrigerator and allow the yogurt to drain for several days. The liquid that accumulates at the bottom is the whey. The strained yogurt can now be used for all sorts of things, such as the Mango Shrikhand on page 109. Use the whey to make the fermented recipes in this section.

FERMENTING EQUIPMENT

Although fermenting generally doesn't require a lot of fancy equipment, there are a few items that will make the fermenting experience easier and more consistent.

The *Lactobacillus* bacteria prefer an *anaerobic* environment, which means without air. As the bacteria break down the sugars and starches, carbon dioxide gas is created, which builds up pressure inside the container. Ideally, the gas should be able to escape without allowing in outside air. This can be accomplished a couple ways. One is by opening the jar or bottle daily and burping it, which lets in a small amount of outside air. Another is by adding an airlock. An airlock is a device often used in brewing beer that forces escaping air though a water barrier, keeping the outside air out.

Although no special containers are required for fermentation, there are a few general guidelines. It is best to use glass or ceramic containers, especially for smaller quantities. I usually use glass because it is easy to see what is going on inside the ferment, the container is easy to clean and sterilize, and it has a relatively small footprint on my kitchen counter.

Lacto-fermentation is an anaerobic process, so most other bacteria that could contaminate the ferment would find the environment inhospitable. For this reason, it is usually good enough to thoroughly clean the fermenting container without needing to take the extra step of sterilizing it. For those who own a dishwasher, running the container through a dishwasher cycle is enough to make the container sterile. For those who do not own a dishwasher, like me, using a boiling water bath for the container will suffice.

ANTIQUE CROCKS FOR FERMENTING

While it may be nostalgic to use great-grandmother's crock for fermentation, I would advise against it, unless the composition of the crock glaze and material is known. Older and foreign crocks may have glazing that is either not intended for food contact or may contain lead, which can leach out during the fermentation process.

FERMENTED KETCHUP

Although technically ketchup doesn't need fermentation, it is a good way to add some probiotics to a common condiment.

INGREDIENTS

2 cans (6 ounces, or 170 g, each) of tomato paste

3 tablespoons (60 g) honey

3 tablespoons (45 ml) apple cider vinegar

2 tablespoons (28 ml) whey

1/4 teaspoon onion powder

1/2 teaspoon salt

1/8 teaspoon black pepper

1/8 teaspoon allspice

Clean pint (475 ml) jar

Canning lid or lid with airlock

Yield: 1 pint (475 ml)

1. Combine all the ingredients in a pint-size (475 ml) canning jar, tasting and adjusting the seasonings as needed. Cover with an airlock or regular lid.

2. Allow the homemade ketchup to sit out at room temperature for 2 to 3 days. If using a regular lid, open the jar every day or so to release the gases. This is not necessary if an airlock is used.

3. Store the ketchup in the refrigerator for another 3 days before enjoying. Keeps for several weeks.

FERMENTED HONEY GARLIC

Fermenting garlic in honey is one of the easiest ferments to do! The fermented garlic mellows and loses some of its pungency, making it a treat to eat. I like to eat a clove or two of the fermented garlic when I feel a cold coming on. Both the honey and the garlic work beautifully for all sorts of dishes, such as the Honey Avocado Dressing on page 121.

I usually make several batches when garlic is plentiful at my local farmers' market. Keep in mind that the number of garlic bulbs and the amount of honey used in this recipe will depend on the size of the garlic cloves.

INGREDIENTS

3 to 5 bulbs of garlic

Approximately 1 cup (340 g) raw honey

Clean pint (475 ml) jar with lid

Yield: 1 pint (475 ml)

1. Peel the garlic cloves and lightly crush them.

2. Fill a pint jar (475 ml) about three-fourths full of garlic and add enough honey to cover while allowing enough head space in the jar for the ferment to bubble, at least 1 to 2 inches (2.5 to 5 cm). Screw the lid on the jar and let it rest on your counter for 1 month.

3. Each day, burp the jar by removing the lid and releasing the built-up air. After 1 month, store in the refrigerator.

FERMENTED HONEY CRANBERRIES

Cranberries or other fruits can also be fermented in honey. Cranberries are grown in my home state of Wisconsin and are usually available fresh each autumn. Other fruits can be used, although I would suggest trying something on the tart side, such as lemons, to counter the sweetness of the honey. Also, I usually make this recipe without spices, but some people add ginger and cinnamon to the ferment.

INGREDIENTS

1 bag (12 ounces, or 340 g) of fresh cranberries

Zest of one orange

Honey to cover, approximately 12 ounces, or 340 g

Strainer

Food processor

Clean quart (950 ml) canning jar with lid

Yield: 1 quart (950 ml)

1. Rinse and sort the cranberries and then lightly pulse the berries in a food processor. The goal is the break them open, not puree them.

2. Add the berries and orange zest to a quart (950 ml) canning jar. Pour the honey over the cranberries and slowly fill up the jar, stopping about 1 to 2 inches (2.5 to 5 cm) from the top.

3. Close the jar and place the jar in a warm, dark place. Turn the jar daily for 1 to 2 weeks until the honey thins and then leave the cranberries to ferment for another 4 to 6 weeks. Store in a cool area.

SPICED VARIATION

If spicing is desired, in step 2, also add 2 to 3 whole cloves, 1 cinnamon stick, and 1 knob of ginger, grated or sliced.

FERMENTED PROBIOTIC HONEY BERRY SODA

This is a naturally fermented soda that is loaded with probiotics, which give the drink its fizz and foam. The recipe is very adaptable depending on what berries are abundant and in season. Although fresh is preferred, frozen berries will work as well. This can also be adapted to other fruit; however, the amount of fruit may need to be adjusted depending on the amount of juice that is released.

INGREDIENTS

5 cups (1.2 L) water

5 cups (weight will vary) berries (crushed)

¾ cup (170 g) honey

½ cup (120 ml) fresh whey (see Straining Yogurt for Whey, page 127)

Additional water to taste

Large saucepan

Thermometer

Strainer or sieve

Clean ½-gallon (1.9 L) glass canning jar with air-lock lid

Wooden spoon

Clean flip-top bottles

Yield: Makes approximately 1½ quarts (1.4 L)

1. In a saucepan, gently simmer the water and berries for approximately 30 minutes. Allow the mixture to cool to about 100°F (38°C).

2. Strain the berry liquid through a sieve into the prepared fermenting jar. Add the honey to the jar, mixing to dissolve it completely. Add the whey and additional water to taste. The mixture will be quite sweet, but much of that sweetness will be used up during the fermentation.

3. Seal the jar with an air-lock lid and leave in a warm spot on the counter for approximately 3 days. Check for fizz and tartness. Fermentation can take up to 1 week or more depending on temperature during the fermentation and the strength of the whey. The warmer the room and the longer the ferment, the fizzier and tart the soda will be.

4. Once it has reached the preferred tartness and fizziness, transfer the soda to the flip-top bottles and refrigerate to slow the fermentation until the soda can be consumed. The soda is usually best when consumed within 2 weeks.

TEPACHE

Tepache is one of Mexico's most popular street drinks, and it is made from fermented pineapple rinds. The final product is a lightly alcoholic, spice-infused glass of fizzy, slightly funky pineapple juice.

This fermented beverage relies on the yeast and bacteria that are naturally present on the pineapple for the fermentation process. To make this, I usually include all the rind and the core and about half of the fruit.

INGREDIENTS

1/2 of a pineapple cut into chunks (Leave the skin on.)

1/2 cup (170 g) dark honey

4 cups (950 ml) water

2 whole cloves

2 tamarind pods

1 cinnamon stick

Knife and cutting board

Clean 1/2 gallon (1.9 L) glass jar

Wooden spoon

Cotton cloth or towel

Strainer

Yield: A little more than 1 quart (950 ml)

1. Wash the pineapple and cut into chunks.

2. Mix the honey and water in the 1/2-gallon (1.9 L) jar until it is completely dissolved.

3. Add the pineapple chunks to the jar and cover with a cotton cloth or towel. Set the jar aside in a cool, dry place away from direct sunlight and let it ferment for 3 to 4 days. It will become cloudy and develop a harmless white foam that can be skimmed off.

4. Strain the finished tepache into a pitcher and refrigerate until well chilled. Serve over ice. This is best consumed within a few days of straining.

BEVERAGES

Although there are lots of ways to use honey in beverages, such as mead or beer, I leave that to the experts and focus on beverages that are easy to make and very tasty.

BASIC HONEY SYRUP
(WITH VARIATIONS)

INGREDIENTS

1/2 cup (170 g) honey	
1/2 cup (120 ml) water	
Medium saucepan	
Wooden spoon	
Yield: 1 cup (235 ml)	

This recipe is the honey equivalent of a sugar-based simple syrup. Simple syrups are handy to keep around for adding sweetness to all sorts of drinks. Why create a simple syrup instead of adding honey directly? Honey on its own is a little too viscous to mix well with other cold ingredients. Honey in hot tea is fine, but honey in an ice-filled cocktail shaker won't mix in nicely. The recipe below is a starting point, but see the sidebar for some interesting additions that can make ordinary drinks extraordinary.

1. Warm the honey and water over medium heat until the honey is completely dissolved and the mixture is homogenous. Do not boil.

2. Let cool completely before using. It can be stored in the refrigerator for up to 2 weeks.

HONEY SYRUP VARIATIONS

Add any of the following to the basic recipe above to make a customized blend.

Ginger: 1/4 to 1 inch (6 mm to 1 cm) thinly sliced

Lemongrass: 1 stalk (core only)

Lavender: 1 tablespoon (2 g) fresh or dried lavender flowers

Dried chili pepper: 1/2 teaspoon, crushed

Vanilla: 1 split vanilla bean

Cardamom: 2 pods, crushed

Rosemary: 3 sprigs

Thyme: 10 sprigs

GINGER ALE

Spicy, sweet, fizzy ginger ale is a favorite of mine any time of year.

INGREDIENTS

2 tablespoons (28 ml) strong ginger honey simple syrup

6 ounces (175 ml) sparkling water

Ice

Twist of lime peel

Cocktail glass

Cocktail stir stick

Yield: Makes 1 serving (8 ounces, or 235 ml)

1. Pour the syrup and sparkling water over ice.

2. Stir gently to combine.

3. Add the lime peel and enjoy.

MANDARIN FIZ

I love the combination of citrus and raspberry.

INGREDIENTS

1/2 cup (120 ml) fresh mandarin or tangerine juice

1/2 teaspoon lemon juice

2 tablespoons (28 ml) basic honey simple syrup

1/2 cup (120 ml) raspberry sparkling water

Ice

Handful of fresh raspberries for garnish

Cocktail glass

Cocktail stir stick

Yield: Makes 1 serving (8 ounces, or 235 ml)

1. Pour all the ingredients over ice.

2. Stir gently to combine.

3. Garnish with raspberries.

CUCUMBER LEMONGRASS COCKTAIL

Lemongrass can be a bit overpowering by itself, but when combined with cucumber, the perfect balance is achieved.

1. Juice ¹/₂ pound (225 g) of cucumbers (or more if needed) in a juicer to yield ³/₄ cup (175 ml) of cucumber juice.

2. Pour the lemongrass honey simple syrup, cucumber juice, and vodka or gin over ice.

3. Stir gently to combine.

4. Garnish with cucumber spear.

INGREDIENTS

³/₄ cup (175 ml) cucumber juice (approximately ¹/₂ pound [225 g] unpeeled cucumbers) and a cucumber spear for garnish

2 tablespoons (28 ml) lemongrass honey simple syrup

1 shot (1.5 ounces, or 42 ml) of vodka or gin

Ice

Juicer or blender

Cocktail glass

Cocktail stir stick

Yield: Makes 1 serving (8 ounces, or 235 ml)

TIP: CUCUMBER JUICE

I use a juicer to make my cucumber juice; however, it can also be made using a blender or food processor. Simply puree the cucumber in the blender and strain the mixture through muslin or several layers of cheesecloth.

APRICOT CARDAMOM COCKTAIL

This cocktail is the perfect balance of sweet, tangy, spicy, bitter, and herbal. A nonalcoholic version of this drink can also be made by mixing with sparkling water.

INGREDIENTS

3 ounces (90 ml) apricot nectar

2 tablespoons (28 ml) cardamom honey simple syrup

1/2 tablespoon lavender honey simple syrup

Splash of grapefruit juice

1 shot (1.5 ounces, or 42 ml) of brandy

Ice

Cocktail glass

Cocktail stir stick

Yield: Makes 1 serving (8 ounces, or 235 ml)

1. Pour all the ingredients over ice.

2. Stir gently to combine.

TEQUILA HONEY COCKTAIL

For me, tequila is the epitome of summer, and I love sitting outside on the patio with friends, sipping margaritas. This cocktail is a pleasant alternative to the standard margarita.

INGREDIENTS

2 ounces (60 ml) tequila

3 tablespoons (45 ml) basic honey syrup (or try a honey syrup variation, such as cardamom)

1 1/2 tablespoons (23 ml) fresh lemon juice

Ice

2 dashes of Angostura bitters

Lemon peel twist for garnish

Cocktail shaker

Cocktail glass

Yield: Makes 1 serving (4 ounces, or 120 ml)

1. Add the tequila, honey syrup, and lemon juice to a shaker with ice and shake until chilled.

2. Pour into a cocktail glass and add 2 dashes of the bitters.

3. Garnish with a lemon peel.

LITHUANIAN HONEY SPIRITS

This Lithuanian spirit is also known as *Krupnikas*. All the spices meld to create this sweet, fragrant blend that is full of complex flavors and aromas. Although technically drinkable in a couple weeks, over time, the flavors continue to improve, so having some patience is key.

INGREDIENTS

- 2 1/4 cups (765 g) honey
- 1 quart (950 ml) water
- 8 whole cloves
- 3 cinnamon sticks
- 10 cardamom pods, cracked
- 1/2 of a whole nutmeg, cracked
- 5 whole allspice, cracked
- 1 1/2 teaspoons black peppercorns
- 1 teaspoon fennel seed
- 3-inch (7.5 cm) ginger root, cut into thick slices
- Zest of 1 orange, peel only, no pith
- Zest of 1/2 of a lemon, peel only, no pith
- 1 vanilla bean, split and scraped
- 1 bottle (750 ml) 190 proof grain alcohol
- Large pot
- Wooden spoon
- Strainer
- Bottles with tops, enough to hold 2 quarts (1.9 L)
- Yield: Makes approximately 2 quarts (1.9 L)

Make a batch right after honey harvest so that some will be ready for the holiday gift-giving season.

1. In a large pot, bring the honey and water to a simmer. Skim off any foam that surfaces.

2. Add all other ingredients except the grain alcohol. Simmer uncovered for 30 minutes.

3. Turn off the heat and add the grain alcohol to the still-hot mixture, stirring to combine. Strain the mixture.

4. Pour into clean, sterile bottles and set aside for at least 2 weeks, longer if possible.

HOLIDAY SPIRIT

This makes a great gift over the holidays. I like to keep a larger bottle on hand for personal use and fill some smaller bottles with the remaining liquor for last-minute gifts.

ELDERBERRY TONIC

Although this really tasty beverage is technically a liqueur, I call it a tonic because during cold and flu season, this has been a godsend. Elderberries have been used for centuries to treat the flu. Elderberry bushes grow wild here in Wisconsin, so I don't have to go too far to find some, but in a pinch, dried berries can also be used. Also, I like to use grain alcohol because there is water in the recipe as well. This ensures that the mixture is preserved properly and doesn't require refrigeration.

INGREDIENTS

2 cups (290 g) fresh elderberries

3 cups (700 ml) water

1 cup (340 g) honey

1 bottle (750 ml) pure grain alcohol, vodka, or brandy

Medium saucepan

Potato masher

Strainer

Bottles with tops, enough to hold 1 quart (950 ml)

Yield: Makes approximately 1 quart (950 ml)

1. Place the elderberries and water in a saucepan. Smash the berries with a potato masher to release the juices. Bring to a boil and allow to cool.

2. Stir in the honey and the alcohol.

3. Pour into clean, sterile bottles and allow to age for at least 1 month.

ELDERBERRY CAUTION

Elderberries are edible when fully ripe, but the leaves, unripe berries, and all other parts of the plant are poisonous. The best way to separate the berries from the stems is to use a fork and "comb" the berries off the stems.

TURMERIC HONEY SUPER BOOSTER

This powerhouse is a trio of super foods. Turmeric is great at fighting inflammation and bacteria. Its active ingredient, curcumin, combats inflammation and swelling in the tissues of the sinuses, which makes it a saving grace for sufferers of hay fever, asthma, and bronchitis as well as seasonal allergies.

INGREDIENTS

¼ cup (85 g) raw honey

1 teaspoon lemon zest

1 tablespoon (7 g) ground turmeric

2 tablespoons (28 ml) raw unfiltered apple cider vinegar

Whisk

Small bowl

Airtight container

Yield: Makes ½ cup (115 ml)

Honey in its raw, unfiltered, unheated state is full of minerals, vitamins, enzymes, and powerful antioxidants. It has healing antibacterial, antiviral, and antifungal properties. It also has pollen grains in it, making it a low-dose pollen therapy.

Organic apple cider vinegar helps to boost immunity. It contains living enzymes that work to fight off infection and bring your body back into balance.

1. Whisk all the ingredients together in a small bowl until smooth. Pour into an airtight container and refrigerate for up to 1 week.

2. To use, simply add 1 tablespoon (15 ml) to some warm water and drink.

APPENDIX A:
SUGAR–HONEY CONVERSION

Converting favorite recipes that call for sugar to honey is not a simple ratio. It depends on what type of recipe it is. Even mild honeys have more sweetening power than sugar. As a general rule, substitute about ¾ cup (255 g) honey for 1 cup (200 g) white sugar. Here are a few things to keep in mind:

To achieve an end result that is as close as possible to the recipe made with white sugar, use the lightest and mildest honey you can find. The darker, stronger honeys could possibly change or overpower some of the other flavors in the recipe.

If the recipe contains a liquid, reduce the amount of liquid slightly. Honey adds/retains more moisture than sugar.

When using honey in baked goods, reduce the oven temperature by about 25°F (14°C). Recipes baked with honey will brown faster in the oven.

Honey is also heavier, denser, and wetter than sugar. Leavening agents may have a harder time rising. Increasing the quantity of leavening agents slightly will counteract this.

APPENDIX B:
USING PERCENTAGES IN RECIPES

All of my recipes are written using percentages. The advantage to doing this is that the recipe can be easily scaled up or down. The disadvantage is that a little math is needed in order to convert the equation into a usable recipe. Here's how to do that.

Here is an example of a recipe:

Shea Butter	31%
Sweet Almond Oil	31%
Beeswax	31%
Flavor Oil	7%

Choose a unit of measure with which to work. Metric units, such as grams or milliliters work best, as the math is more straightforwrad. They can be converted back to English units at the end of the process.

Decide the desired size of your batch. For this example, we will use a 10 ounce batch. That is equal to 283 grams.

Determine the percentage of each portion of the recipe.

Multiply the number by the percent. Example:
[283 (grams) x 0.31(%) = 87.73 (grams)]

Round to the desired precision
[87.73 rounded to the nearest whole number = 88]

Finalize the recipe.

Using the sample recipe, it now looks like this:

Shea Butter	88 grams
Sweet Almond Oil	88 grams
Beeswax	88 grams
Flavor Oil	20 grams

APPENDIX C:
LYE CALCULATIONS

These are some of my favorite online soap-making lye calculators. They provide varying degrees of information and differ in ease of use for the novice soap maker.

www.thesage.com/calcs/LyeCalc.html

soapcalc.net/

www.brambleberry.com/pages/lye-calculator.aspx

Although I highly recommend using an online lye calculator, there may be times when it is necessary to calculate the amount of lye needed by hand.

In this appendix, I have listed all the oils I use in this book and their corresponding fatty acid profiles and values needed to do the lye calculations. The SAP values are given in ranges, as the actual values vary from one batch of oil to the next and from one supplier to the next. If the actual SAP number is known for a particular oil, that number should be used. The SAP number is expressed as the number of milligrams of KOH (potassium hydroxide) required to saponify 1 gram of fat. A Certificate of Analysis (COA) generally shows this number.

In order to use this number to calculate the lye needed, the SAP value needs to be converted into a universal multiplier. There is a difference in numbers because the molecular weights of KOH and NaOH (sodium hydroxide) are not the same.

To simply things, I have already done the math using an average SAP value for particular oils. I have included those converted values next to the actual SAP value for both sodium hydroxide (NaOH) and potassium hydroxide (KOH).

In order to calculate the lye needed, using the converted SAP value for the desired oil, simply multiply the desired weight of oil by the converted SAP number.

For example, to make the Bar Soap with Honey recipe, these are the calculations:

For liquid soap (KOH) = the SAP value ÷ 1000

For solid soap (NaOH) = the SAP value ÷ 1402.50

This number is not the same as the amount of lye shown in the recipe. That's because the number calculated above will convert all of the oil into soap. However, because most of the SAP numbers used are just averages of an acceptable range of lipids for a particular oil, it is recommended that a buffer is calculated in, about 2 to 3 percent. On top of that, this soap would probably be very drying, so it would be helpful to have some oil remaining in the soap to act as a moisturizer, let's say about 5 percent, bringing the total buffer to 7 percent. This 7 percent number is called the *superfat*. I generally work with 6 to 7 percent superfat in my bar soap recipes.

So, with the above example, we take:

	NaOH SAP	%	Ounces	Lye needed
Coconut	0.180	35.4%	9.31	1.68
Rice bran	0.128	11.4%	3.00	0.38
Olive	0.135	30.4%	8.00	1.08
Avocado	0.133	7.6%	2.00	0.27
Shea	0.128	10.3%	2.71	0.35
Castor	0.128	4.9%	1.29	0.16
			Total lye	3.92 ounces

3.92 oz NaOH x (100% − 7%) = 3.6 oz NaOH needed

APPENDIX D:
BODY CARE INGREDIENT GUIDE

Ingredients INCI (International Nomenclature of Cosmetic Ingredients) names appear below in italics.

BUTTERS AND OILS

AVOCADO OIL

Persea gratissima (Avocado) Oil

DESCRIPTION:

This is a medium-weight oil with a shelf life of about 1 year. Avocado oil is easily absorbed into deep tissue, and with its wonderfully emollient properties, it is ideal for mature skin. It also helps to relieve the dryness and itching of psoriasis and eczema. It is also said to help with skin regeneration.

FATTY ACID PROFILE:

Caprylic Acid	0.0%
Capric Acid	0.0%
Lauric Acid	0.0%
Myristic Acid	15.2%
Palmitic Acid	5.5%
Stearic Acid	0.1%
Oleic Acid	62.0%
Linoleic Acid	16.0%
Linolenic Acid	0.0%
Ricinoleic Acid	0.0%

SAP VALUE RANGE:

177–198

CONVERTED SAP:

KOH / NaOH	.188 / .133

CASTOR OIL

Ricinus Communis (Castor) Seed Oil

DESCRIPTION:

This is one of the thickest and heaviest oils available. It has a shelf life of 5 years. On skin, it is a humectant and helps to retain moisture levels. In soap, it makes for a softer bar but improves lather when used in small quantities. It can also make the soap a bit translucent.

FATTY ACID PROFILE:

Caprylic Acid	0.0%
Capric Acid	0.0%
Lauric Acid	0.0%
Myristic Acid	0.0%
Palmitic Acid	2.0%
Stearic Acid	1.0%
Oleic Acid	7.0%
Linoleic Acid	3.0%
Linolenic Acid	0.0%
Ricinoleic Acid	87.0%

SAP VALUE RANGE:

176–186

CONVERTED SAP:

KOH / NaOH	.181 / .128

COCOA BUTTER

Theobroma Cacao (Cocoa) Seed Butter

DESCRIPTION:

This is a hard butter with a melting temperature of 100°F (38°C). It has a relatively long shelf life of 2 to 5 years. It also contains polyphenols and phyosterols and some skin softening Vitamin E. Cocoa butter provides an occlusive layer on our skin.

FATTY ACID PROFILE:

Caprylic Acid	0.0%
Capric Acid	0.0%
Lauric Acid	0.0%
Myristic Acid	0.5%
Palmitic Acid	25.0%
Stearic Acid	35.0%
Oleic Acid	32.0%
Linoleic Acid	3.0%
Linolenic Acid	0.0%
Ricinoleic Acid	0.0%

SAP VALUE RANGE:

188–200

CONVERTED SAP:

KOH / NaOH	.194 / .137

COCONUT OIL

Cocos Nucifera (Coconut) Oil

DESCRIPTION:

This is solid at temperatures below 76°F (24°C). This is a medium-weight oil with a shelf life of about 2 to 4 years. On skin, coconut oil contains some very potent antioxidants that can help with skin aging and help repair damage caused by light and radiation. It softens, moisturizes, and soothes chapped and itchy skin. In soap, coconut oil creates loads of big bubbles. Although it makes a hard bar, it melts very quickly in water, making it a great soap for hard water conditions. It is also extremely cleansing, but in high proportion, it can be very drying to the skin.

FATTY ACID PROFILE:

Caprylic Acid	7.1%
Capric Acid	6.0%
Lauric Acid	48.0%
Myristic Acid	16.5%
Palmitic Acid	8.0%
Stearic Acid	3.8%
Oleic Acid	5.0%
Linoleic Acid	2.5%
Linolenic Acid	0.0%
Ricinoleic Acid	0.0%

SAP VALUE RANGE:

250–264

CONVERTED SAP:

KOH / NaOH	.257 / .180

GRAPESEED OIL

Vitis vinifera (Grape) Seed Oil

DESCRIPTION:

This is another great lightweight oil. It has a relatively short shelf life of 6 months to 1 year. On skin, it can help to retain moisture and reduce inflammation and itchiness. In soap, I recommend that it is used sparingly because of its short shelf life.

FATTY ACID PROFILE:

Caprylic Acid	0.0%
Capric Acid	0.0%
Lauric Acid	0.0%
Myristic Acid	0.0%
Palmitic Acid	8.0%
Stearic Acid	4.0%
Oleic Acid	18.0%
Linoleic Acid	68.0%
Linolenic Acid	0.0%
Ricinoleic Acid	0.0%

SAP VALUE RANGE:

185–200

CONVERTED SAP:

KOH / NaOH	.193 / .123

JOJOBA OIL

Simmondsia Chinensis (Jojoba) Seed Oil

DESCRIPTION:

Technically, this is not an oil but rather a liquid wax ester. It is lightweight and has a long shelf life of at least 2 years. It is great for hair and skin because it mixes with sebum, which allows it to be washed away. It is an excellent emollient. In soap, it will add moisturizing and conditioning properties as well as extending the shelf life of the soap.

FATTY ACID PROFILE:

Caprylic Acid	0.0%
Capric Acid	0.0%
Lauric Acid	0.0%
Myristic Acid	0.0%
Palmitic Acid	0.0%
Palmitoleic Acid	2.0%
Stearic Acid	0.0%
Oleic Acid	0.0%
Linoleic Acid	0.0%
Linolenic Acid	0.0%
Ricinoleic Acid	0.0%

SAP VALUE RANGE:

90–95

CONVERTED SAP:

KOH / NaOH	.093 / .069

MANGO BUTTER

Mangifera Indica (Mango) Seed Butter

DESCRIPTION:

This is a medium-hard butter with a melting temperature of 86°F to 98.6°F (30°C to 37°C) and a slight sweet scent. It has a shelf life of 2 to 3 years. It has emollient properties, high oxidative ability, and wound healing and regenerative abilities.

FATTY ACID PROFILE:

Caprylic Acid	0.0%
Capric Acid	0.0%
Lauric Acid	0.0%
Myristic Acid	0.0%
Palmitic Acid	9.0%
Stearic Acid	38.0%
Oleic Acid	41.0%
Linoleic Acid	6.0%
Linolenic Acid	0.0%
Ricinoleic Acid	0.0%

SAP VALUE RANGE:

183–198

CONVERTED SAP:

KOH / NaOH	.191 / .128

OLIVE OIL

Olea Europaea (Olive) Fruit Oil

DESCRIPTION:

This a medium-weight oil with a shelf life of about 1 year. It is known for its skin moisturizing and nourishing properties. It makes an extremely gentle soap, suitable for sensitive skin types. Used by itself in soap, it yields a lather that is creamy rather than bubbly and while still fresh, the lather may actually feel a bit slimy. That sliminess goes away after a 6-month or more cure time.

FATTY ACID PROFILE:

Caprylic Acid	0.0%
Capric Acid	0.0%
Lauric Acid	0.0%
Myristic Acid	0.0%
Palmitic Acid	14.0%
Stearic Acid	3.0%
Oleic Acid	71.0%
Linoleic Acid	10.0%
Linolenic Acid	1.0%
Ricinoleic Acid	0.0%

SAP VALUE RANGE:

184–196

CONVERTED SAP:

KOH / NaOH	.190 / .135

RICE BRAN OIL

Oryza Sativa (Rice) Bran Oil

DESCRIPTION:

Rice Bran oil is a medium-weight oil with a shelf life of 6 months to 1 year. It is high in vitamin E and fatty acids that make skin soft and improve elasticity. It also helps with cell regeneration. In soap, it behaves similarly to olive oil, but I find it adds silkiness to the soap and makes a slightly harder, longer lasting bar.

FATTY ACID PROFILE:

Caprylic Acid	0.0%
Capric Acid	0.0%
Lauric Acid	0.0%
Myristic Acid	0.0%
Palmitic Acid	15.0%
Stearic Acid	2.0%
Oleic Acid	42.0%
Linoleic Acid	39.0%
Linolenic Acid	1.0%
Ricinoleic Acid	0.0%

SAP VALUE RANGE:

185–195

CONVERTED SAP:

KOH / NaOH	.185 / .128

SHEA BUTTER

Butyrospermum parkii (Shea Butter) Fruit

DESCRIPTION:

This is a soft butter with a melting temperature of 97°F to 100°F (36°C to 38°C). It has a shelf life of 2 to 3 years. Shea penetrates deep into the skin, rejuvenating damaged cells and restoring elasticity and tone. In soap, it has a larger than average proportion of fats that are not saponifiable, meaning your soap will actually retain some moisturizing qualities.

FATTY ACID PROFILE:

Caprylic Acid	0.0%
Capric Acid	0.0%
Lauric Acid	0.0%
Myristic Acid	0.0%
Palmitic Acid	5.0%
Stearic Acid	40.0%
Oleic Acid	47.0%
Linoleic Acid	5.0%
Linolenic Acid	0.0%
Ricinoleic Acid	0.0%

SAP VALUE RANGE:

170–190

CONVERTED SAP:

KOH / NaOH	.180 / .128

SWEET ALMOND OIL

Prunus dulcis (Almond) oil

DESCRIPTION:

This is a lightweight oil with a shelf life of 1 year. This emollient oil softens and nourishes the skin. It is often used for massage. In soap, it creates a conditioning, stable lather.

FATTY ACID PROFILE:

Caprylic Acid	0.0%
Capric Acid	0.0%
Lauric Acid	0.0%
Myristic Acid	0.0%
Palmitic Acid	0.0%
Stearic Acid	6.5%
Oleic Acid	69.4%
Linoleic Acid	17.4%
Linolenic Acid	0.0%
Ricinoleic Acid	0.0%

SAP VALUE RANGE:

190–200

CONVERTED SAP:

KOH / NaOH	.195 / .136

MISCELLANEOUS INGREDIENTS

CITRIC ACID

Citric acid

DESCRIPTION:

Citric acid is primarily used to adjust the acidity of a formulation. It is naturally occurring in citric fruits and juices.

E-WAX NF

E-wax NF

DESCRIPTION:

E-wax is an all-in-one, easy to use, emulsifier to combine water and oil into a lotion or cream. NF means that it conforms to the National Formulary guidelines.

GLYCERIN

Glycerin

DESCRIPTION:

Glycerin is a humectant, drawing moisture from the air. It also forms a protective layer that helps prevent moisture loss.

LANOLIN

Lanolin

DESCRIPTION:

Technically, lanolin is a wax. It is the yellow, extremely viscous substance sheep produce that helps them to shed water from their coats. Most lanolin is a by-product of wool processing, as it needs to be removed prior to dyeing or weaving. For skin, it is a wonderful protectant that creates a semi-permeable layer to keep the elements out and the moisture in.

PRESERVATIVE (OPTIPHEN)

Phenoxyethanol (and) caprylyl glycol

DESCRIPTION:

Optiphen is a broad-spectrum paraben-free preservative that can be used to preserve emulsified products that contain water, such as lotions and creams. There are other preservatives, but this is one of my favorites.

ROSEMARY OLEORESIN EXTRACT (ROE)

Rosmarinus Officinalis (Rosemary) Leaf Extract

DESCRIPTION:

Rosemary oleoresin, or ROE as it is often abbreviated, is added to oils as an antioxidant. It helps to keep oils from oxidizing and becoming rancid and it prolongs the shelf life of a product. This is not a preservative.

STEARIC ACID

Stearic acid

DESCRIPTION:

See Fatty Acid Glossary

VITAMIN E OIL (T-50 OR TOCOPHEROLS)

Tocopherol

DESCRIPTION:

T-50 is an all-natural blend of mixed tocopherols (Vitamin E) and edible vegetable oils. It functions as an excellent antioxidant for oils or products that contain oil. It helps delay oxidation and will help to extend the shelf life of oils and oil-based products. It works a bit differently than the rosemary oleoresin, so I often use both.

HERBS

ARNICA

Arnica montana
Arnica is known for its natural anti-inflammatory properties and its ability to reduce swelling. Arnica salve is a muscle workhorse. It is wonderful for sprains, strains, and pulled ligaments or muscles! It also helps bruised areas heal.

CALENDULA BLOSSOMS

Calendula officinalis
Anti-inflammatory and antimicrobial, calendula blossoms are good for minor cuts and burns. It promotes healing by aiding clotting.

CHAMOMILE, GERMAN

Matricaria chamomilla
German chamomile contains the flavonoids apigenin, luteolin, and quercetin, and the volatile oils alpha-bisabolol and matricin, which help to make chamomiles anti-inflammatory, antispasmodic, and antioxidant. It is also known for being calming, soothing, and healing.

CHAMOMILE, ROMAL

Anthemis nobilis
See German Chamomile—Roman chamomile has a very similar effect on skin.

CHICKWEED LEAF

Stellaria media
Chickweed leaf is a great soother of itchy and sore skin. It helps to cool and relieve inflamed areas.

COMFREY LEAF & ROOT

Symphytum officialis
The high allantoin content in comfrey is the star of this herb. It encourages the renewal of skin cells, the strengthening of skin tissues, and helps to heal wounds as well as reducing the visible effects of aging by bringing back elasticity to the skin. It is also a mild astringent, which makes comfrey great for sunburn, ulcers, and sores. It is so powerful that it should not to be used on deep cuts because it can trap infection by healing the top layer of skin faster than deeper layers.

LAVENDER

Lavandula angustifolia
Lavender is a superstar for burns and encouraging wound healing. It has also been found to be effective against the principal bacteria involved in acne. Lavender has been known to soothe insect bites and bee stings as well.

MARSHMALLOW ROOT

Althaea officinalis
Marshmallow root soothes skin and promotes wound healing.

PLAINTAIN LEAF

Plantago major
With astringent, antibacterial, anti-inflammatory, and anti-itch properties, plantain leaf has a long history of being used for problem skin. It contains salicylic acid, which helps to reduce acne breakouts. It is also great for bee stings. Just pick a leaf, mash it up a bit, and apply as a poultice to the sting.

ROSEMARY

Rosmarinus officinalis
Rosemary removes excess oil from skin without causing excess dryness. It has regenerating and stimulating effects that help rejuvenate skin and give it a more youthful glow. It helps to restore elasticity and firmness to the skin and encourages skin cell turnover and renewal.

SELF-HEAL

Prunella vulgaris
Self-heal has astringent, antibacterial, and antiseptic properties, and it helps to reduce pain and promotes healing of cuts and scrapes. It can even be used on skin rashes.

ST. JOHN'S WORT

Hypericum perforatum
St. John's wort helps to support skin elasticity, soothes redness and irritation, and works wonders on burns, wounds, insect bites, and other skin irritations.

THYME

Hypericum perforatum
Thyme's natural antimicrobial, antibacterial, and anti-inflammatory properties help to remove excess oil and dirt from pores and reduce acne outbreaks. Thyme contains powerful antioxidants that can help protect skin from premature aging.

APPENDIX E:
FATTY ACID GLOSSARY

The *C* number after the fatty acid name corresponds to the number of carbon molecules in the fatty acid.

CAPRYLIC ACID (C8)

Caprylic acid is used for its emollient, hydrating, and antifungal properties.

CAPRIC ACID (C10)

Often found with caprylic and lauric acids in coconut and palm kernel oil, capric acid is used to moisturize and replenish skin.

LAURIC ACID (C12)

Lauric acid is the most abundant fatty acid in coconut and palm kernel oil. It is reputed to have anti-inflammatory, antiviral, and antimicrobial properties.

MYRISTIC ACID (C14)

Myristic acid is a saturated acid naturally found in coconut oil and palm kernel oil. It is used in cosmetics for its cleansing, smoothing, and protective properties.

PALMITIC ACID (C16)

Palmitic acid helps to reinforce skin barrier function; it is a very good emollient.

PALMITOLEIC ACID (C16:1)

Palmitoleic acid is a building block in our skin that prevents burns, wounds, and skin scratches. The most active microbial in human sebum, palmitoleic acid is used to treat damaged skin and mucous membranes.

STEARIC ACID (C18)

Stearic acid helps improve moisture retention and increases flexibility of the skin and skin damage repair.

OLEIC ACID (C18:1)

Oleic acid is very moisturizing and helps skin cells regenerate quickly. It is absorbed very well by the skin and also acts as an anti-inflammatory.

LINOLEIC ACID (C18:2)

Linoleic acid helps to improve skin's barrier function and helps to soothe itchy, dry skin. It also acts as an anti-inflammatory and a moisture retainer.

LINOLENIC ACID (C18:3)

Known for its hydrating power, linolenic acid is an anti-inflammatory that soothes redness and irritation.

RICINOLEIC ACID (C18:3, N-9)

Only found in castor oil, ricinoleic acid has analgesic and anti-inflammatory effects.

GADOLEIC ACID (C20:1)

Gadoleic acid acts to prevent transdermal water loss. It serves as an occlusive barrier without being sticky or overly greasy. In fact, most users describe it as having a smooth finish, which makes it a favorite for use by massage therapists.

ERUCIC ACID (C22:1)

Erucic acid is used in cosmetic products as an emollient because it provides a protective layer for skin.

APPENDIX F:
BODY CARE SUPPLIER INDEX

Below is a list of suppliers I use and recommend. Some companies are listed in multiple categories. I have chosen to list suppliers this way to make it easier to find a specific ingredient. All of the suppliers listed below are ones I use in my own business. Many of the suppliers listed below also carry items from other categories that may be very good, but are untested by me.

Also, it is intentional on my part to include only businesses that seem to have withstood the test of time, as I hope this list will be a reference for a number of years. There may be newer and smaller suppliers out there that have exceptional quality and fair pricing.

BEESWAX

BEEHIVE ALCHEMY
www.beehivealchemy.com/
Address: 2115 N 56th St
Milwaukee, WI 53208
Tel: 414-403-1993

DADANT & SONS, INC
www.dadant.com/
Address: 51 South 2nd Street
Hamilton, IL 62341
Tel: 217-847-3324
Tel: 888-922-1293

LYE (SODIUM HYDROXIDE AND POTASSIUM HYDROXIDE)

BOYER CORPORATION
www.boyercorporation.com/
Address: P.O. Box 10
La Grange, IL 60525
Tel: 708-352-2553
Tel: 800-323-3040

WHOLESALE SUPPLIES PLUS
www.wholesalesuppliesplus.com/
Address: 7820 E. Pleasant Valley Rd
Independence, OH 44131
Tel: 800-359-0944

BASE OILS AND BUTTERS

SOAPER'S CHOICE (A DIVISION OF COLUMBUS FOODS COMPANY (CFC, INC))
www.soaperschoice.com/
Address: 30 E. Oakton Street
Des Plaines, IL 60018
Tel: 847-257-8946
Tel: 800-322-6457 ext: 8946

BRAMBLE BERRY
www.brambleberry.com/
Address: 301 W. Holly St
Bellingham, WA 98225
Tel: 360-676-1030
Tel: 877-627-7883

WHOLESALE SUPPLIES PLUS
www.wholesalesuppliesplus.com/
Address: 7820 E. Pleasant Valley Rd
Independence, OH 44131
Tel: 800-359-0944

CAMDEN-GREY ESSENTIAL OILS, INC
www.camdengrey.com/
Address: 3567 NW 82 Avenue
Doral, FL 33122 US
Tel: 305-500-9630

NEW DIRECTIONS AROMATICS
www.newdirectionsaromatics.com/
Address: 6781 Columbus Road
Mississauga, Ontario, L5T 2G9
Canada
Tel: US: 800-246-7817
Tel: CAN: 877-255-7692

ESSENTIAL OILS AND FRAGRANCE OILS

AROMA HAVEN/ RUSTIC ESCENTUALS, LLC

rusticescentuals.com/
Address: 1050 Canaan Road
Roebuck, SC 29376
Tel: 864-384-5331

NEW DIRECTIONS AROMATICS

www.newdirectionsaromatics.com/
Address: 6781 Columbus Road
Mississauga, Ontario, L5T 2G9
Canada
Tel: US: 800-246-7817
Tel: CAN: 877-255-7692

WHITE LOTUS AROMATICS

www.whitelotusaromatics.com/
Address: 332 Carriage Dr.
Sequim, WA 98382
Tel: 360-683-0137

CAMDEN-GREY ESSENTIAL OILS, INC

www.camdengrey.com/
Address: 3567 NW 82 Avenue
Doral, FL 33122
Tel: 305-500-9630

CONTAINERS AND PACKAGING

SKS BOTTLES & PACKAGING, INC

www.sks-bottle.com/
Address: 2600 7th Avenue
Building 60 West
Watervliet, NY 12189
Tel: 518-880-6980 ext: 1

WHOLESALE SUPPLIES PLUS

www.wholesalesuppliesplus.com/
Address: 7820 E. Pleasant Valley Rd
Independence, OH 44131
Tel: 800-359-0944

BOTTLES AND MORE

www.bottlesandmore.com/
Address: 859 Stillwater Rd #1
West Sacramento, CA 95605
Tel: 916-995-4557

HERBS AND BOTANICALS

MONTEREY BAY SPICE COMPANY, INC

www.herbco.com
Address: 241 Walker St
Watsonville, CA 95076
Tel: 831-722-3400
Tel: 800-500-6148

ROCKY MOUNTAIN SPICE COMPANY

www.rockymountainspice.com
Address: 3850 Nome St
Denver, CO 80239
Tel: 303-308-8066
Tel: 888-568-4430

MOUNTAIN ROSE HERBS

www.mountainroseherbs.com
Address: PO Box 50220
Eugene, OR 97405
Tel: 541-741-7307
Tel: 800-879-3337

ENCAUSTIC AND BATIK SUPPLIES

DICK BLICK ART MATERIALS
www.dickblick.com
Tel: 1-800-828-4548

THE EARTH PIGMENTS COMPANY, LLC
www.earthpigments.com
Tel: 520-682-8928

SPECIALTY COSMETIC INGREDIENTS

THE HERBARIE AT STONEY HILL FARM, INC.
www.theherbarie.com/
Address: Prosperity, SC 29127
Tel: 803-364-9979

LOTIONCRAFTER
www.lotioncrafter.com/
Address: 48 Hope Ln
Eastsound, WA 98245
Tel: 360-376-8008

INGREDIENTS TO DIE FOR
www.ingredientstodiefor.com/
Address: 11110 Metric Blvd, STE D
Austin, TX 78758
Tel: 512-535-2711

MAJESTIC MOUNTAIN SAGE INC
https://www.thesage.com/
Address: 2490 South 1350 West
Nibley, Utah 84321
Tel: 435-755-0863

WHOLESALE SUPPLIES PLUS
www.wholesalesuppliesplus.com/
Address: 7820 E. Pleasant Valley Rd
Independence, OH 44131
Tel: 800-359-0944

BRAMBLE BERRY
www.brambleberry.com/
Address: 2138 Humboldt St.
Bellingham, WA 98225
Tel: 360-676-1030
Tel: 877-627-7883

FOREIGN SUPPLIERS

VOYAGEUR SOAP & CANDLE COMPANY LTD
www.voyageursoapandcandle.com/
Address: #14 – 19257 Enterprise Way
Surrey, BC V3S 6J8
Tel: 604-514-0632
Tel: 800-758-7773

GRACEFRUIT LIMTED
www.gracefruit.com/
Address: 146 Glasgow Road, Longcroft,
Stirlingshire, FK4 1QL, UK
Tel: Telephone: 01324 841353

THE SOAP KITCHEN
www.thesoapkitchen.co.uk/
Address: Unit 8, Coddsdown Industrial
Park
Clovelly Road, Bideford
Devon, EX39 3DX UK
Tel: +44(0) 1237 420 872

SIDNEY ESSENTIAL OIL COMPANY
www.seoc.com.au/
Address: 68 Melverton Drive
Hallam, VIC 3803 Australia
Tel: 02 9565 2828

NEW DIRECTIONS AUSTRALIA
www.newdirections.com.au/
Address: 47 Carrington Road,
Marrickville, Sydney, NSW 2204 Australia
Tel: (612) 8577 5999
Tel: 1800 637 697

KOSMETISCHE ROHSTOFFE
www.kosmetische-rohstoffe.de/
Address: Alberti Habermann • Am
Grenzberg 34 • 63654 Büdingen
Germany
Tel: (0 60 42) 17 66

ACKNOWLEDGMENTS

I would like to thank Karl for taking me along on his beekeeping journey. It is his passion for bees that continues to inspire me on a daily basis. I would also like to thank the team at Quarry Books who have placed their faith in me not once, but twice, and my photographer Kat Schleicher whose photographs make the ordinary look extraordinary. I am grateful for their understanding of my need for flexibility to still operate my business Beehive Alchemy and write this book.

ABOUT THE AUTHOR

PETRA AHNERT is the creative force behind Beehive Alchemy (beehivealchemy.com), a growing artisan soap, body care, and home goods business. After a serendipitous meeting with a beekeeper at the library in 2004, Petra soon had a couple hives of her own and extra honey to sell. Of course, with the honey comes beeswax, too, so she sought out ways to make good use of the wax she had on hand. Since that time, Petra has first looked to her own life and products she uses to replace them with items she could make herself with the honey and beeswax she had on hand. Eventually, those products made their way into the product lineup for Beehive Alchemy.

INDEX

ALSO AVAILABLE FROM QUARRY BOOKS

Beeswax Alchemy
978-1-59253-979-6

The Backyard Beekeeper, 4th Edition
978-1-63159-332-1

The Backyard Beekeeper's Honey Handbook
978-1-59253-474-6

The Benevolent Bee
978-1-63159-286-7